SpringerBriefs in Mathematics

SpringerBriefs in Mathematics showcases expositions in all areas of mathematics and applied mathematics. Manuscripts presenting new results or a single new result in a classical field, new field, or an emerging topic, applications, or bridges between new results and already published works, are encouraged. The series is intended for mathematicians and applied mathematicians.

More information about this series at http://www.springer.com/series/10030

Mickaël D. Chekroun · Honghu Liu
Shouhong Wang

Approximation of Stochastic Invariant Manifolds

Stochastic Manifolds for Nonlinear SPDEs I

Springer

Mickaël D. Chekroun
University of California
Los Angeles, CA
USA

Shouhong Wang
Indiana University
Bloomington, IN
USA

Honghu Liu
University of California
Los Angeles, CA
USA

ISSN 2191-8198 ISSN 2191-8201 (electronic)
SpringerBriefs in Mathematics
ISBN 978-3-319-12495-7 ISBN 978-3-319-12496-4 (eBook)
DOI 10.1007/978-3-319-12496-4

Library of Congress Control Number: 2014956373

Mathematics Subject Classification: 37L65, 37D10, 37L25, 35B42, 37L10, 37L55, 60H15, 35R60, 34F05, 34G20, 37L05

Springer Cham Heidelberg New York Dordrecht London

Printed on acid-free paper

Springer International Publishing AG Switzerland is part of Springer Science+Business Media (www.springer.com)

To our families

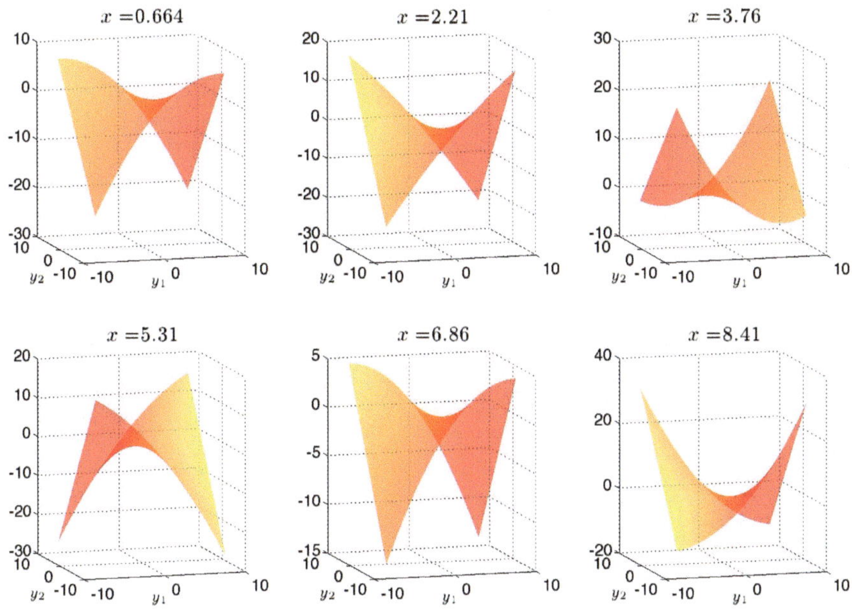

An approximation of a stochastic hyperbolic manifold associated with a stochastic Burgers-type equation: x-disintegration over the leading pair of stable/unstable modes, for a given realization of the noise. The calculations are based on the formulas derived in this monograph; see Chap. 7.

Preface

The intriguing figure from the previous page presents a visualization of the type of manifolds this monograph is concerned with. First, obtained as a graph over some chosen linearized spatial modes from a nonlinear stochastic partial differential equation (SPDE), such a manifold—and its geometric properties—depends naturally on the spatial variable. Second, the shape of this manifold is nonlinear as a fingerprint of the nonlinear effects conveyed by the nonlinear SPDE. Third, this manifold is stochastic, in other words, it depends on time for a given realization of the noise, meaning that "mutations" of the shape—such as changes in the (Gaussian) curvature—can occur as time flows. Manifolds that share these three features and more importantly for which the dynamics of the underlying nonlinear SPDE is either attracted to or meanders around, are the manifolds of interest in this two-volume series.

In this respect, a *pathwise* approach from the theory of random dynamical systems (RDS) is adopted in both volumes. Volume I deals with *approximation* of stochastic manifolds that are invariant for the SPDE dynamics, while Volume II [37] deals with stochastic manifolds that may not be invariant nor attractive and that can still capture essential features of the dynamics through appropriate parameterizations of the small spatial scales by the large ones. The small-scale parameterizations proposed in Volume II are articulated around a new concept of stochastic manifolds, namely the stochastic *parameterizing manifolds* (PMs).

An important common feature is shared by the (pathwise) approximation formulas derived in Volume I and the (pathwise) parameterization techniques introduced in Volume II: both are characterized as pullback limits from backward–forward systems which are only partially coupled, facilitating the calculation of such limits, either analytically or numerically.

The aforementioned pullback limits arise under the form of Lyapunov–Perron integrals which are useful for the rigorous treatment of the problem of approximation to the leading order,[1] of important stochastic manifolds such as—but not

[1] With respect to the nonlinearity involved in the SPDE at hand.

limited to—stochastic critical manifolds[2] built as random graphs over a fixed number of critical modes which lose their stability as a control parameter varies.

In this respect, Chaps. 6 and 7 contain explicit formulas for the leading-order Taylor approximation of stochastic (local) critical manifolds or more general stochastic hyperbolic manifolds. The pullback characterization of these formulas provides a useful interpretation of the corresponding approximating manifolds which gives rise to a simple framework that allows us, furthermore, to unify the previous approximation approaches of stochastic invariant manifolds, as discussed in Volume II; see [37, Sect. 4.1].

To help the reader appreciate this unification as well as the more prospective material presented in Volume II, we took the freedom to include a self-contained (short) survey on the theory of existence and attraction of one-parameter families of stochastic invariant manifolds in Chaps. 4 and 5.

Los Angeles, September 2014 — Mickaël D. Chekroun
Los Angeles — Honghu Liu
Bloomington — Shouhong Wang

[2] Including stochastic manifolds such as the center or center-unstable manifolds.

Acknowledgments

Preliminary versions of this work were presented by the authors at the AIMS Special Session on "Advances in Classical and Geophysical Fluid Dynamics," held at the Ninth AIMS Conference on Dynamical Systems, Differential Equations and Applications in Orlando, July 2012; at the "workshop on Random Dynamical Systems," held at the Institute of Mathematics and Applications in October 2012; at the "Lunch Seminars" held at the Center for Computational and Applied Mathematics, Purdue University, in November 2012; at the "Applied Mathematics Seminar," held at the University of Illinois at Chicago in March 2013; at AMS Special Session on "Partial Differential Equations from Fluid Mechanics," held at University of Louisville in October 2013; at the "PDE seminar," held at Indiana University in October 2013; and at the Alpine Summer School on Dynamics, Stochastics, and Predictability of the Climate System, Valle d'Aosta, Italy in June 2014.

We would like to thank all those who helped in the realization of this book through encouragement, advice, or scientific exchanges: Jerry Bona, Jerome Darbon, Jinqiao Duan, Michael Ghil, Nathan Glatt-Holtz, Dmitri Kondrashov, Tian Ma, James McWilliams, David Neelin, James Robinson, Jean Roux, Eric Simonnet, Roger Temam, Yohann Tendero, and Kevin Zumbrun. More particularly, the authors are grateful to Michael Ghil for his constant support and interest in this work, over the years. MDC is grateful to James McWilliams and David Neelin for stimulating discussions on the closure problem of turbulence and stochastic parameterizations; and to Michael Ghil and Roger Temam for stimulating discussions on the slow and the "fuzzy" manifold. MDC thanks also Jinqiao Duan for the reference [103], and for discussions on stochastic invariant manifolds.

We would like to express also our thanks to Ute McCrory (Springer) for her invaluable patience and support during the preparation of this book.

MDC and HL are supported by the National Science Foundation grant DMS-1049253 and Office of Naval Research grant N00014-12-1-0911. SW is supported in part by National Science Foundation grants DMS-1211218 and DMS-1049114, and by Office of Naval Research grant N00014-11-1-0404. SW is most grateful to

Wen Masters and Reza Malek-Madani for their support and appreciation for his work over the years.

Last but not the least, we would like to express all of our gratitude to our wives and children for their unflinching love and support during the realization of this project.

Contents

Acronyms

AIM	Approximate inertial manifold
MDS	Metric dynamical system
NR	Non-resonance
OU	Ornstein–Uhlenbeck process
PDE	Partial differential equation
PES	Principle of exchange of stabilities
PM	(Stochastic) Parameterizing manifold
RDE	Random differential equation
RDS	Random dynamical system
RPDE	Random partial differential equation
SDE	Stochastic differential equation
SPDE	Stochastic partial differential equation

Chapter 1
General Introduction

The theory of invariant manifolds for deterministic dynamical systems has been an active research field for a long time, and is now a very well-developed theory; see, e.g., [6, 7, 8, 9, 10, 30, 43, 44, 52, 62, 74, 79, 80, 88, 89, 92, 94, 105, 117, 118, 119, 121, 132, 143, 147, 148, 149, 150, 152, 153, 154]. Over the past two decades, several important results on random invariant manifolds for stochastically perturbed ordinary as well as partial differential equations (PDEs) have been obtained; these results often extend those found in the deterministic setting; see, e.g., [1, 2, 3, 12, 20, 22, 25, 26, 29, 51, 57, 65, 66, 113, 123, 124, 128, 142]. Even so, the stochastic theory is still much less complete than its deterministic counterpart. For instance the reduction problem of a stochastic partial differential equation (SPDE) to its corresponding stochastic invariant manifolds has been much less studied and only a few works in that direction are available [18, 29, 40, 67, 103, 145, 155].

The practical aspects of the reduction problem of a deterministic dynamical system to its corresponding (local) center, center-unstable or unstable manifolds have been well investigated in various finite- and infinite-dimensional settings; see, e.g., [13, 23, 30, 59, 70, 74, 75, 90, 91, 92, 98, 100, 108, 111, 117, 121, 133, 134]. In the stochastic but finite-dimensional context, extensions of center-manifold reduction techniques have been first investigated in [21, 22], and completed by procedures which consist of deriving simultaneously both normal forms and center manifold reductions of stochastic differential equations (SDEs) in [2, 3, 128, 156]; see [1, Sect. 8.4.5]. We mention also [53] for prior works on stochastic normal forms.

For stochastic systems of interest for applications, and SPDEs in particular, the existence problem of stochastic invariant manifolds can be reasonably solved only *locally* in practice. This is due to conditions inherent to the global theory of stochastic invariant manifolds which can be fulfilled only locally via a standard *cut-off procedure* of the nonlinear terms when the latter are not globally Lipschitz; see Chap. 5. Such conditions involve typically a gap in the spectrum of the linear part which has to be large enough in comparison with the variation of the nonlinear terms, expressed under the form of various *spectral gap conditions* in the literature; see Theorems 4.1 and 4.3.

M.D. Chekroun et al., *Approximation of Stochastic Invariant Manifolds*,
SpringerBriefs in Mathematics, DOI 10.1007/978-3-319-12496-4_1

In situations where the basic state loses its stability as a control parameter λ crosses a critical value λ_c, such local stochastic invariant manifolds are typically obtained (for λ sufficiently close to λ_c) as random graphs over a neighborhood $\mathcal{V}$ of this basic state, contained in the subspace spanned by the modes which lose their stability, called hereafter the critical modes. As a consequence, any reduction procedure based on these manifolds can make sense only for λ sufficiently close to λ_c for which the amplitudes $a_\lambda(t)$ of the critical modes remain sufficiently small so that $a_\lambda(t) \in \mathcal{V}$. Such a condition can be satisfied for all t, in the case of deterministic autonomous systems [90, 117], or even in the case of non-autonomous ones with appropriate time-bounded variations of the vector field [5, 59, 135]. However in a stochastic context, due to large excursions of the solutions caused by the (white) noise, this condition is expected to be unavoidably violated even when the magnitude of the noise σ is small; and is expected to be violated more frequently (as time t flows) when σ gets large. We refer to [48, 115] for results about *large deviation principle* for SPDEs driven by multiplicative noise.

It is the purpose of Volume II [37] to propose a new type of stochastic manifolds that are *pathwise global* objects[1] which are able to circumvent this difficulty while allowing us to derive efficient reduced systems which are able to achieve good modeling performances of the SPDE dynamics projected onto the resolved (e.g., critical) modes, even away from the critical value where the amplitude of these modes gets large. These manifolds are not invariant in general but provide an approximate parameterization of the unresolved variables by the resolved ones, which improve in mean square error—over any (sufficiently large) finite time interval—the partial knowledge of the full SPDE solution u when compared with the one obtained from $u_{\mathfrak{c}} := P_{\mathfrak{c}} u$ alone,[2] for a given realization of the noise; see [37, Definition 4.1]. Such manifolds are naturally named *stochastic parameterizing manifolds* (*PMs*) hereafter. A computable criterion coined as the *parameterization defect* and inherent to [37, Definition 4.1], makes furthermore possible the practical verification of whether a given stochastic manifold constitutes a stochastic PM; see [37, Sect. 6.1].

Interestingly, stochastic PMs are not subject to a spectral gap condition such as encountered in the classical theory of stochastic invariant manifolds as revisited in Chaps. 4 and 5 of the current volume. Instead, certain stochastic PMs can be determined under weaker *non-resonance conditions* in the self-adjoint case: For any given set of resolved modes for which their self-interactions through the nonlinear terms do not vanish when projected against a given unresolved mode e_n, it is required that some specific linear combinations of the corresponding eigenvalues dominate the eigenvalue associated with e_n; see e.g., (NR)-condition in Chap. 7 and (NR2)-condition in [37, Sect. 7.3]. No constraints are thus imposed on the Lipschitz constant, explaining why the stochastic PM-theory can overcome the locality issue caused by cut-off arguments.

---

[1] Defined as graphs of random continuous functions $h(\xi, \omega)$ defined for each realization ω over the *whole* subspace $\mathcal{H}^{\mathfrak{c}}$ spanned by the resolved modes, i.e., for all $\xi \in \mathcal{H}^{\mathfrak{c}}$, where $\mathcal{H}^{\mathfrak{c}}$ is typically the subspace spanned by the first few eigenmodes with low wavenumbers.

[2] Here $P_{\mathfrak{c}} u$ denotes the projection of u onto the resolved modes.

It is also shown in [37] that a certain type of stochastic PMs coincides actually with the leading-order approximation of, for instance, the local stochastic center or unstable manifolds when restricted to the appropriate neighborhood $\mathcal{V}$ of the basic state; see [37, Sect. 4.1]. The classical theory can thus be reconnected with the important difference that stochastic PMs are global objects and allow thus for large amplitudes.

To further clarify the relationships between stochastic PMs and the classical theory, this monograph is devoted to the derivation of leading-order Taylor approximation for certain types of local stochastic invariant manifolds associated with a broad class of SPDEs driven by linear multiplicative white noise. In particular, approximation formulas—extending those of [120, Theorem A.1.1]—for local stochastic critical manifolds[3] are derived in Chap. 6 (see Theorem 6.1 and Corollary 6.1) and for local stochastic hyperbolic manifolds, in Chap. 7. Classical theorems about existence and attraction properties of stochastic invariant manifolds are also revisited in Chap. 4 to make the expository as much self-contained as possible; see Theorem 4.3 and Corollary 4.3. We invite the interested reader to consult Chap. 2 to get a more detailed overview of the main results of the current volume.

The notion of (stochastic) PMs such as introduced in [37], raises naturally the question of its relation with the theory of (stochastic) approximate inertial manifolds (AIMs). The latter has been thoroughly investigated in the deterministic literature, and numerous candidates to a substitute of inertial manifolds have been introduced in that purpose; see, e.g., [60, 61, 62, 78, 81, 129, 151]. In all the cases, the idea was to relax the requirements of the inertial manifold theory so that the longterm dynamics can be at least described in some approximate sense, still by some finite-dimensional manifold. Particular efforts have been devoted to developing efficient methods to determine in practice such manifolds. This has led in particular to the so-called nonlinear Galerkin methods; see e.g., [24, 49, 63, 78, 80, 99, 101, 102, 122, 129, 147, 151]. Approximate inertial manifolds have also been considered in a stochastic context [50], but contrary to the deterministic case, only very few algorithms are available to compute stochastic AIMs in practice; see, e.g., [103].

Such a context motivates the important problem examined in [37] concerning the practical determination of stochastic PMs. A general approach is introduced in that respect in [37, Chap. 4]. This approach can be viewed as the cornerstone between the two volumes of this book. It consists of obtaining candidates of such PMs as pullback limits of the unresolved variables $\mathbf{z}$ as modeled by auxiliary *backward-forward systems*. The key idea consists here of representing the modes with high wavenumbers as a pullback limit depending on some approximations of the time history of the modes with low wavenumbers. Such an idea is not new and has been used in the context of 2D-turbulence [69], with the essential difference that the pullback limits considered Volume II [37], are associated with backward-forward systems that are *partially coupled* in the sense that only the (past values) of the

[3]Defined as graphs over a neighborhood $\mathcal{V}$ contained in the subspace spanned by the critical modes, for λ sufficiently close to λ_c.

resolved variables $\mathbf{y}$ force the equations of the unresolved variables $\mathbf{z}$, without any feedback in the dynamics of the former.

Due to this partial coupling, given a realization ω of the noise, the equations for the resolved variables can be first integrated backward over an interval $[-T, 0]$ $(T > 0)$ from $\xi \in \mathcal{H}^{\mathfrak{c}}$, and equations for the unresolved variables are then integrated forward over $[0, T]$. From this operation, the state of the unresolved variables $\mathbf{z}$ at time s $(s > 0)$ is thus conditioned on ξ denoted as $\mathbf{z}[\xi]$, and depends on the state of the resolved variables as well as on the noise at time $s - T$. The unresolved variables $\mathbf{z}$ as modeled by such a system is thus a function of the history of both the resolved variable $\mathbf{y}$ and the noise.

The pullback limit of $\mathbf{z}[\xi]$ obtained as $T \to \infty$, when it exists, gives generally access to a parameterizing manifold function $h(\xi, \omega)$; but of course, the design of appropriate backward-forward systems is essential for such an operation to be successful. Different strategies are introduced in [37, Sect. 4.3] in that respect. Conditions under which such systems give access to stochastic PMs are identified for the stochastic context in [37, Sect. 4.4] (see Theorems 4.2 and 4.3 therein), as well as for the deterministic one in [37, Sect. 4.5]. Interestingly, as for certain AIMs [78], relations with time analyticity properties of the PDE solutions are involved in order that the pullback limit associated with certain backward-forward systems, provides a stochastic PM; see [37, Theorem 4.4]. The obtention of stochastic PMs via such pullback limits are further analyzed on a stochastic Burgers-type equation in [37, Chaps. 6 and 7].

It is worthwhile to mention that such a pullback characterization of stochastic PMs, constitutes an approach more appealing, from a numerical viewpoint, when compared to methods of approximation of *stochastic inertial manifolds* directly rooted in the work of [57] which are also based on backward-forward systems, but this time, *fully coupled*; see [103].[4] We mention also that from a more standard point of view, the complementary pullback characterization of (local) approximating manifolds presented in [37, Sect. 4.1] provides a novel interpretation of such objects in terms of flows. The framework set up in this way allows us, furthermore, to unify the previous approximation approaches from the literature [18, 29, 40, 155], and we mention also that these features are not limited to the stochastic setting as discussed in [37, Chap. 4].

Finally, two *global* reduction procedures based on stochastic PMs obtained as pullback limits from the auxiliary systems of [37, Sect. 4.3], are respectively presented in [37, Chap. 5] when analytic expression of such PMs are available, and in [37, Chap. 7] in the general case. In the case where analytic expressions are not available, a numerical procedure is described in [37, Sect. 7.2] (and in [37, Sect. 7.3]) to determine "on the fly" the reduced random vector field along a trajectory $\xi(t, \omega)$ generated by the latter as the time is advanced. The cornerstone in this case, is (again)

[4]Furthermore, we mention that our particular choice of multiplicative noise allows us to consider—via the cohomology approach (see Sect. 3.3)—transformed versions (ω by ω) of our backward-forward systems (such as [37, system (4.3)]) so that we do not have to deal with adaptiveness issues which arise in solving more general stochastic equations backward in time [103].

the pullback characterization of the appropriate stochastic PMs, which allows us to update the reduced vector field once $\xi(t, \omega)$ is known at a particular time instance t.

In all the cases, a main feature of the resulting PM-based reduced systems, comes from the *interactions between the stochastic and nonlinear effects* which are shown to contribute to the emergence of noise-induced memory effects (see [37, Lemma 5.1] and [37, Sect. 7.3]) conveyed by the stochastic parameterizing manifolds such as built [37, Chap. 4]. These memory effects are shown to play an essential role in the derivation of reduced models able to describe with a good accuracy the main dynamical features of the amplitudes of the (resolved) modes contained in $\mathscr{H}^{\mathfrak{c}}$; see [37, Chaps. 6 and 7]. Interestingly, the PM-based stochastic reduction procedure can be seen as an alternative to the (stochastic) nonlinear Galerkin method where the approximate stochastic inertial manifolds (AIMs) used therein [50] are substituted here by the parameterizing manifolds introduced above.

Among the differences with the AIM approach, the PM approach seeks for manifolds which allows to provide—in a *mean square sense*—simple modeling error estimates for the evolution of $u_{\mathfrak{c}}$, over any finite (sufficiently large) time interval; see [37, Proposition 5.1] and see also [34, Theorem 1 and Corollary 1]. This modeling error is controlled by the product of three terms: the energy of the unresolved modes (i.e., the unknown information), the nonlinear effects related to the size of the global random attractor, and the parameterization defect of the stochastic PM employed in the reduction. Interestingly there are cases where parameterization defect can be easily assessed by the theory. For instance, when the trivial steady state is unstable, a stochastic inertial manifold (when it exists) is shown to always constitute a stochastic PM; see [37, Theorem 4.1]. The corresponding parameterization defect Q decays then to zero, and the parameterization of the small spatial scales by the large ones becomes asymptotically "exact" in that case; see again [37, Theorem 4.1]. Complementarily, Theorems 4.2, 4.3, and 4.4 in [37] deal with situations where a stochastic inertial manifold is not known to exist, and provide theoretical estimates of the parameterization defect of various stochastic PMs.

In more precise mathematical terms, the aforementioned PM-based reduced models are low-dimensional SDEs arising typically with random coefficients which convey *extrinsic memory effects* [86, 87] expressed in terms of decay of correlations (see [37, Lemma 5.1]), making the stochastic reduced equations genuinely non-Markovian [86]. These random coefficients involve the history of the noise path and exponentially decaying terms depending in the self-adjoint case on the gap between some linear combinations of the eigenvalues associated with the low modes and the eigenvalues associated with the high modes. These gaps correspond exactly to those appearing in the non-resonance condition (NR) ensuring the pullback limit—associated with the first backward-forward system introduced in [37, Sect. 4.3]—to exist. In that case, the memory terms emerge from the nonlinear leading-order interactions between the low modes,[5] embedded in the "noise bath".

[5] As projected onto the high modes.

Extrinsic memory effects of different type have been encountered in reduction strategies of finite-dimensional SDEs to random center manifolds; see, e.g., [21]. Extrinsic memory effects also arise in procedures which consist of deriving simultaneously both normal forms and center manifold reductions of SDEs; see for instance [2, 3, 128] and [1, Sect. 8.4.5]. In such a two-in-one strategy, anticipating terms may arise—as integrals involving the future of the noise path—in both the corresponding random change of coordinates and the resulting normal form.

In [83, 116, 136], pursuing the works of [2, 3], reduced stochastic equations involving also extrinsic memory terms have been derived mainly in the context of the stochastic slow manifold; see also [17]. By seeking for a random change of variables, which typically involves repeated stochastic convolutions, reduced equations (different from those derived in [37, Chap. 5]) are obtained to model the dynamics of the slow variables. These reduced equations are also non-Markovian but require a special care in their derivation to push the anticipative terms (arising in such an approach) to higher order albeit not eliminating them [83, 136].

As a comparison, our reduction strategy is naturally associated with the theory of (stochastic) parameterizing manifolds of Volume II [37], and in particular it does not require the existence of a stochastic slow (or inertial) manifold. Our approach prevents furthermore the emergence of anticipative terms to any order in the corresponding reduced SDEs. Memory terms of more elaborated structures than described in [37, Lemma 5.1] (see, e.g. [37, (7.36) or (7.40)]) can also arise in our stochastic reduced equations built from stochastic PMs defined as pullback limits associated with the *multilayer backward-forward systems* introduced in [37, Sect. 4.3]. As illustrated for the stochastic Burgers-type equation analyzed in [37, Chap. 7], such a multilayer backward-forward system conveys typically a *hierarchy of memory terms* obtained via repeated compositions of functions involving integrals depending on the past of the noise path driving the SPDE. Such a hierarchy arises with higher-order terms resulting from a "*matrioshka*" *of nonlinear self-interactions* between the low modes, as well as with a sequence of non-resonance conditions, both of increasing complexity.

As application, it is shown in [37, Sect. 7.5] that such elaborated memory terms and higher-order terms may turn out to become particularly relevant for a faithful reproduction—from a PM-based reduced model—of statistical features[6] of the dynamics on the low modes. Such a situation is shown to occur far from the criticality, when the nonlinear cross-interactions between the high and low modes, as well as the self-interactions among the high modes, contribute significantly to the dynamics on the low modes.

It is furthermore shown in [37, Sect. 7.5] that non-trivial noise-induced phenomena—such as large excursions of the low-mode amplitudes—can be reproduced with high-accuracy from a PM-based reduced model, even when the amount of noise is significant and the separation of time scales is weak. Such a success is attributed to

[6] Such as the autocorrelation and probability density functions.

the ability of the underlying stochastic PM to capture, for a given realization and as time flows, the noise-driven transfer of energy to the small spatial scales through the nonlinear term.

Finally, we mention that the proposed framework to deal with the problem of approximate parameterizations of the small spatial scales by the large ones for SPDEs, has been intentionally articulated for the case of linear multiplicative noise (also known as parameter noise [16]), in order to present the main ideas in a simple stochastic context. We emphasize that this framework is not limited to that case and actually extends to SPDEs driven by multidimensional noise, either multiplicative or additive; we refer to [35] for SPDEs driven by additive noise forcing finitely many modes. Similarly, deterministic PMs can be defined and efficiently computed in the deterministic setting as briefly discussed in [37, Sect. 4.5] and further investigated in [34] for the design of low-dimensional suboptimal controllers of nonlinear parabolic PDEs.

Chapter 2
Stochastic Invariant Manifolds: Background and Main Contributions

The focus of this first volume is the derivation of leading-order approximations of stochastic invariant manifolds such as stochastic center manifolds by extending, to a stochastic context, the techniques described in [117, Chap. 3] and [120, Appendix A]; see Chap. 6. New properties of approximating manifolds described in terms of pullback characterization will be reported in Volume II [37, Sect. 4.1].[1] The framework set up in this way allows us, furthermore, to unify the previous approximation approaches from the literature [18, 29, 40, 103]. These features are not limited to the stochastic setting as pointed out in Volume II [37, Sect. 4.1].

In that respect and motivated by the study of stochastic bifurcations or more general phase transitions arising in nonlinear SPDEs[2] [56, 127], we first revisit in Chaps. 4 and 5 the existence and smoothness properties (Theorems 4.1, 4.2 and 5.1)—as well as the attraction properties in terms of *almost sure asymptotic completeness* (Theorem 4.3)—of families of global stochastic invariant manifolds parameterized by the noise amplitude σ, and by some control parameter λ. The latter is assumed here to vary in some interval Λ over which a uniform decomposition of the spectrum holds; see (3.11) below. The latter condition implies some uniform (partial)-dichotomy estimates that are satisfied by the linearized stochastic flow about the basic state; see (3.46).

The questions of existence and smoothness are dealt within a framework rooted in the standard Lyapunov-Perron method [19, 109, 114, 131]. The techniques follow those used for instance in [46, 92, 152, 153, 154], from which we propose a treatment adapted to the random setting inspired mainly by the works of [42, 66]. The related existence and smoothness results are essentially known, but are revisited here in order to set up the precise framework and to provide the technical tools on which we

[1] Section 4.1 in Volume II [37] concerns the approximating manifolds considered in this volume. This somewhat unconventional presentation has been adopted here in order to articulate, in a unified way, the *pullback characterization* of such manifolds as well as of the *stochastic parameterizing manifolds* considered in [37].

[2] which is the main purpose of [36].

M.D. Chekroun et al., *Approximation of Stochastic Invariant Manifolds*, SpringerBriefs in Mathematics, DOI 10.1007/978-3-319-12496-4_2

rely to establish the main results of this volume in Chaps. 6 and 7 as well as those presented in Volume II [37, Sect. 4.1].

Our treatment of the *asymptotic completeness problem* is inspired by the work of [46] that we adopt in the stochastic framework, and that consists of reformulating this problem as a fixed point problem under constraints; see (4.21). The latter problem is then recast as an unconstrained fixed point problem associated with a random integral operator (see 4.24) that is solved by means of the *uniform contraction mapping principle* [44, Theorems 2.1 and 2.2]. Various types of attraction properties of random invariant manifolds have been explored in the literature mostly in contexts where the associated SPDEs possess a stable self-adjoint linear part and a bounded and Lipschitz nonlinearity. For instance, in [12, Theorem 3.1], both forward and pullback exponential attractions[3] of the stochastic inertial manifold are established. Asymptotic completeness in some nth-moment has been established in [51, Theorem 2] and [57, Proposition 3.5] (with n being any integer in [51] and $n = 2$ in [57]) for stochastic inertial manifolds associated with certain types of SPDEs with respectively additive and multiplicative noise. Almost sure forward asymptotic completeness for deterministic initial data has been established in [42] for retarded SPDEs with additive noise and a stable self-adjoint linear part, adapting also the work of [46] to a stochastic context. Almost sure pullback asymptotic completeness of stochastic invariant manifolds has also been investigated in [155, Theorem 2.1] for certain type of SPDEs with nonlinearities which do not cause a loss of regularity compared to the ambient space $\mathscr{H}$.[4]

For SPDEs considered in this monograph, which in particular allow for nonlinearities causing a loss of regularity (see Chap. 3), Theorem 4.3 provides conditions under which the stochastic invariant manifolds ensured by Corollary 4.1 are almost surely forward and pullback asymptotically complete with respect to random tempered initial data; see Definition 4.3. In particular the existence of a one-parameter family of global stochastic inertial manifolds is obtained, and it is shown that the constitutive manifolds of this family attract exponentially the dynamics at a uniform rate as λ varies in Λ. The results obtained in Theorem 4.3 and Corollary 4.3 are not restricted to the case of self-adjoint linear operator and include the cases where unstable modes are present. The latter situation is particularly useful to establish in Volume II [37] that stochastic inertial manifold always constitute a stochastic PM; see [37, Theorem 4.1] whose proof relies furthermore on some elements contained in the proof of Theorem 4.3.

In Chap. 5, we present a local theory of stochastic invariant manifolds associated with the global theory described in Chap. 4. The ideas are standard but the material is detailed here again in view of the derivation of the main results regarding the approximation formulas of stochastic critical manifolds (Chap. 6) and local stochastic

[3] The exponential attraction used therein extends in a random context the classical one encountered in the theory of (deterministic) inertial manifold [79].

[4] Namely, $F\colon \mathscr{H} \to \mathscr{H}$, adopting the notations of Chap. 3. Note also that the proof of [155, Theorem 2.1] provided therein is not complete.

hyperbolic invariant manifolds (Chap. 7), as well as the related pullback characterizations discussed in Volume II [37, Sect. 4.1].

Chapters 6 and 7 are devoted to the main results concerning approximating manifolds of local stochastic critical manifolds on the one hand (Chap. 6), and local hyperbolic ones, on the other (Chap. 7). They concern the derivation of new approximation formulas of these (local) stochastic invariant manifolds. More precisely, in Chap. 6, we consider the important case for applications where some leading modes lose (once) their stability as λ varies in Λ, which is formulated as the *principle of exchange of stabilities (PES)*; see condition (6.4). It is shown in Lemma 6.1 that the latter implies the uniform spectrum decomposition assumed in previous sections. This allows us in turn to establish in Proposition 6.1, the existence of a family of *local stochastic critical manifolds* which are built—by relying on Chap. 5—as graphs over some deterministic neighborhood of the origin in the subspace spanned by the critical modes that lose their stability as λ varies.[5] By construction, these manifolds carry nonlinear dynamical information associated with the loss of the linear stability of these critical modes; see [36].

We then derive in Theorem 6.1 and Corollary 6.1, *explicit random approximation formulas* to the leading order of these local stochastic critical manifolds about the origin. These stochastic critical manifolds are built naturally as graphs over a fixed number of critical modes, which lose their stability as λ varies. More precisely, the corresponding approximating manifolds are obtained as graphs—over some λ-independent neighborhood $\mathcal{N}$ of zero in the subspace $\mathcal{H}^{\mathfrak{c}}$ spanned by the critical modes—of the following one-parameter family of random functions:

$$\mathfrak{I}_\lambda(\xi, \omega) = \int_{-\infty}^{0} e^{\sigma(k-1)W_t(\omega)\mathrm{Id}} e^{-tL_\lambda} P_{\mathfrak{s}} F_k(e^{tL_\lambda}\xi)\,\mathrm{d}t, \quad \xi \in \mathcal{N}, \qquad \text{(AF)}$$

where F_k denotes the leading-order nonlinear terms of order k, L_λ the corresponding parameterized linear part, $P_{\mathfrak{s}}$ the projector upon the non-critical modes, and $W_t(\omega)$ the Wiener path associated with the realization ω of the noise with amplitude σ.

It is worth mentioning at this stage that the random approximation formulas such as (AF), contrast with the deterministic ones proposed in [18] and [40] for certain types of SPDEs. In particular, the nonlinearity considered in [18] consists of a bilinear term, $B(u, u)$, while it consists of power nonlinearity, u^k, with $k \geq 2$ in [40]. The error bounds for the approximation of the local random invariant manifold function $h(u, \omega)$ provided in both [18] and [40] are of the same order as $\|u\|$ and are valid with large probability, and for sufficiently small u; see [40, Lemma 4.10] and footnote 3 in [37, Chap. 4] for [18, Theorem 7].

The class of nonlinear SPDEs of type (3.1) considered below contains the SPDEs dealt with in [18, 40] as special cases. In contrast with the deterministic approximation formulas obtained in [18, 40], the approximations derived hereafter are genuine random polynomial functions, which approximate almost surely the local random

[5] See Definition 6.1 for more details.

critical manifolds and provide (random) Taylor approximations of these manifolds to the leading order; see Corollary 6.1. More precisely, a priori error estimates are derived in a general setting which are of order $o(\|u\|_\alpha^k)$ if the nonlinear term, $F(u)$, is such that $\|F(u)\| = O(\|u\|_\alpha^k)$ for some integer $k \geq 2$;[6] see again Theorem 6.1 and Corollary 6.1 for precise statements of these results.

Approximation formulas such as given by (AF) are then extended to the case of stochastic hyperbolic manifolds in Chap. 7, which allows for the low-dimensional subspace $\mathscr{H}^{\mathfrak{c}}$ to contain a combination of critical modes, and modes that remain stable as λ varies in some interval Λ. In that respect, relaxation of the conditions on the spectrum under which the Lyapunov-Perron integral $\mathfrak{I}_\lambda$ exists, are identified. In particular, when L_λ is self-adjoint, it is shown that $\mathfrak{I}_\lambda$ exists if a non-resonance condition (NR) is satisfied: For any given set of resolved modes for which their self-interactions (through the leading-order nonlinear term F_k) do not vanish when projected against an unresolved mode e_n, it is required that some specific linear combinations of the corresponding eigenvalues dominate the eigenvalue associated with e_n.

We turn now to the organization of this first volume. In Chap. 3, we introduce the class of SPDEs considered throughout this monograph and describe the main assumptions among which a uniform decomposition of the spectrum of the linear part constitutes a key ingredient in most of the proofs presented hereafter. We also recall some basic concepts from RDS theory [1], and cast such SPDEs into the RDS framework by a classical random change of variables leading to random partial differential equations (RPDEs). The existence, uniqueness, and measurability properties of classical solutions to such RPDEs are recalled in Proposition 3.1. To make the expository as much self-contained as possible, the proof and some related results concerning the mild solutions to these RPDEs are presented in Appendix A.

In Chap. 4, we revisit the existence and attraction properties of global random/stochastic invariant manifolds within a framework that is suitable for the derivation of certain results regarding the stochastic parameterizing manifolds introduced in Volume II [37]; see, e.g., [37, Theorem 4.1]. We first derive the existence and smoothness of such manifolds for the transformed RPDEs in Theorems 4.1 and 4.2. The corresponding results for the original SPDEs are presented in Corollaries 4.1 and 4.2. Finally, the almost sure forward-and-pullback asymptotic completeness of these manifolds is examined in Theorem 4.3 and Corollary 4.3.

In Chap. 5, we relax the global Lipschitz condition on the nonlinear term, and derive accordingly the existence of local stochastic invariant manifolds for SPDEs; see Theorem 5.1 and Corollary 5.1. Chapter 6 is devoted to the main results of this first volume regarding the approximation formulas of local stochastic critical manifolds for SPDEs, as summarized in Theorem 6.1 and Corollary 6.1. Rigorous error estimates to the leading order are in particular derived. These results are then extended to the case of (local) stochastic hyperbolic manifolds in Chap. 7.

[6] Here $\|\cdot\|_\alpha$ denotes a norm on a space of functions more regular than those of the ambient space $\mathscr{H}$; see Chap. 3.

Chapter 3
Preliminaries

In this chapter, we introduce the functional framework and our standing hypotheses concerning the abstract stochastic evolution equations of type (3.1) below that we will work with. We also recall some basic concepts from the RDS theory [1, 54], and introduce a classical random change of variables [66] which will be used to cast a given parameterized family of SPDEs into the RDS framework.

3.1 Stochastic Evolution Equations

We consider the following nonlinear stochastic evolution equation[1] driven by linear multiplicative white noise in the sense of Stratonovich:

$$\mathrm{d}u = \big(L_\lambda u + F(u)\big)\mathrm{d}t + \sigma u \circ \mathrm{d}W_t. \tag{3.1}$$

Here, $\{L_\lambda\}$ represents a family of linear operators parameterized by a scalar control parameter λ, $F(u)$ accounts for the nonlinear terms, W_t is a two-sided one-dimensional Wiener process, and σ is a positive constant which gives a measure of the "amplitude" of the noise.

Such equations arise in various contexts such as in turbulence theory or non-equilibrium phase transitions [15, 56, 127], in the modeling of randomly fluctuating environment [11] in spatially-extended harvesting models [38, 93, 125, 126, 139, 140], or simply in the modeling of parameter disturbances [16].

We make precise below the functional framework that we will adopt throughout this monograph for the examination and approximation of a natural class of stochastic invariant manifolds associated with Eq. (3.1).

[1]Throughout this monograph, we will often refer to a stochastic evolution equation of type (3.1) as an SPDE.

M.D. Chekroun et al., *Approximation of Stochastic Invariant Manifolds*,
SpringerBriefs in Mathematics, DOI 10.1007/978-3-319-12496-4_3

Assumptions about the operator L_λ. Let $(\mathscr{H}, \|\cdot\|)$ be an infinite-dimensional separable real Hilbert space. First, let us introduce a sectorial operator A on $\mathscr{H}$ [92, Definition 1.3.1] with domain

$$\mathscr{H}_1 := D(A) \subset \mathscr{H}, \tag{3.2}$$

and which has compact resolvent. In particular $\mathscr{H}_1$ is compactly and densely embedded in $\mathscr{H}$. We assume furthermore that $-A$ is stable in the sense that its spectrum satisfies $\operatorname{Re}\sigma(-A) < 0$.

We shall also make use of the fractional powers of A and the associated interpolated spaces between $\mathscr{H}_1$ and $\mathscr{H}$; see, e.g., [92, Sect. 1.4] and [143, Sect. 3.7]. Let $\mathscr{H}_\gamma := D(A^\gamma)$ be such an interpolated space for some $\gamma \in [0, 1]$, endowed with the norm $\|\cdot\|_\gamma$ induced by the inner product $\langle u, v\rangle_\gamma := \langle A^\gamma u, A^\gamma v\rangle_{\mathscr{H}}$; in particular $\mathscr{H}_0 = \mathscr{H}$, $\mathscr{H}_1$ corresponds to $\gamma = 1$, and $\mathscr{H}_1 \subset \mathscr{H}_\gamma \subset \mathscr{H}_0$ for $\gamma \in (0, 1)$. Note that in the sequel, $\langle\cdot,\cdot\rangle$, will be used to denote the inner-product in the ambient Hilbert space $\mathscr{H}$.

Let us introduce now

$$B_\lambda \colon \mathscr{H}_\gamma \to \mathscr{H} \tag{3.3}$$

a parameterized family of bounded linear operators depending continuously on λ, with here $\gamma \in [0, 1)$. In particular, $-B_\lambda A^{-\gamma}$ is bounded on $\mathscr{H}$ and according to [92, Corollary 1.4.5] the operator $-L_\lambda$ is sectorial on $\mathscr{H}$ with domain $\mathscr{H}_1$ where

$$L_\lambda := -A + B_\lambda. \tag{3.4}$$

Note that L_λ has compact resolvent by recalling that $\mathscr{H}_1$ is compactly embedded in $\mathscr{H}$ [71, Proposition II.4.25]. As a consequence, since $L_\lambda \colon \mathscr{H}_1 \to \mathscr{H}$ is a closed operator,[2] we have that for each λ, the spectrum of L_λ, $\sigma(L_\lambda)$, consists only of isolated eigenvalues with finite algebraic multiplicities; see [104, Theorem III-6.29] (see also [71, Corollary IV.1.19]).

Assumptions about the nonlinearity F. For the nonlinearity, we assume that $F \colon \mathscr{H}_\alpha \to \mathscr{H}$ is continuous for some $\alpha \in [0, 1)$ which will be fixed throughout this monograph.[3]

We assume furthermore that[4]

$$F(0) = 0, \tag{3.5}$$

and in the case where F is at least C^1-smooth, the tangent map of F at 0 is assumed to be the null map, i.e.,

---

[2] As a consequence of the sectorial property of $-L_\lambda$.

[3] In particular, nonlinearities including a loss of regularity compared to the ambient space $\mathscr{H}$, are allowed; see, e.g., Volume II [37, Chaps. 6–7] for an illustration.

[4] Near a nontrivial steady state $\overline{u}$ of some deterministic system, one can think u as some deviation from this steady state (subject to noise fluctuations), and Eq. (3.5) is then satisfied in such situations.

$$DF(0) = 0. \tag{3.6}$$

Note that in particular, according to (3.5) the noise term in (3.1) is multiplicative with respect to the basic state; see [1, p. 473] for this terminology.

Moreover, for the results on global random invariant manifolds proved in Chap. 4, F will be assumed to be furthermore globally Lipschitz:

$$\|F(u) - F(v)\| \leq \mathrm{Lip}(F)\|u - v\|_\alpha, \quad \forall u, v \in \mathscr{H}_\alpha, \tag{3.7}$$

where $\mathrm{Lip}(F)$ denotes the smallest positive constant such that (3.7) is true. Other assumptions on F will be specified when needed; see, e.g., Theorem 4.2 and Chap. 6.

The spectrum of L_λ and the uniform spectrum decomposition. Recall that the spectrum $\sigma(L_\lambda)$ consists only of isolated eigenvalues with finite multiplicities. This property combined with the sectorial property of $-L_\lambda$ implies that there are at most finitely many eigenvalues with a given real part. The sectorial property of $-L_\lambda$ also implies that $\mathrm{Re}\,\sigma(L_\lambda)$ is bounded above (see also [71, Theorem II.4.18]). These two properties of $\mathrm{Re}\,\sigma(L_\lambda)$ allow us in turn to label elements in $\sigma(L_\lambda)$ according to the lexicographical order which we will adopt throughout this monograph:

$$\sigma(L_\lambda) = \{\beta_n(\lambda) \mid n \in \mathbb{N}^*\}, \tag{3.8}$$

such that for any $1 \leq n < n'$ we have either

$$\mathrm{Re}\,\beta_n(\lambda) > \mathrm{Re}\,\beta_{n'}(\lambda), \tag{3.9}$$

or

$$\mathrm{Re}\,\beta_n(\lambda) = \mathrm{Re}\,\beta_{n'}(\lambda), \quad \text{and} \quad \mathrm{Im}\,\beta_n(\lambda) \geq \mathrm{Im}\,\beta_{n'}(\lambda). \tag{3.10}$$

Note that we will adopt in this monograph the convention that each eigenvalue, $\beta_n(\lambda)$, is repeated according to its algebraic multiplicity.

We assume throughout this monograph that an open interval Λ can be chosen such that the following *uniform spectrum decomposition* of $\sigma(L_\lambda)$ holds over Λ:

$$\sigma(L_\lambda) = \sigma_{\mathfrak{c}}(L_\lambda) \cup \sigma_{\mathfrak{s}}(L_\lambda), \quad \lambda \in \Lambda, \qquad \text{with} \qquad \eta_{\mathfrak{c}} > \eta_{\mathfrak{s}}, \tag{3.11}$$

where

$$\begin{aligned} \eta_{\mathfrak{c}} &:= \inf_{\lambda\in\Lambda} \inf\{\mathrm{Re}\,\beta(\lambda) \mid \beta(\lambda) \in \sigma_{\mathfrak{c}}(L_\lambda)\}, \\ \eta_{\mathfrak{s}} &:= \sup_{\lambda\in\Lambda} \sup\{\mathrm{Re}\,\beta(\lambda) \mid \beta(\lambda) \in \sigma_{\mathfrak{s}}(L_\lambda)\}, \end{aligned} \tag{3.12}$$

and $\sigma_{\mathfrak{c}}(L_\lambda)$ consists of the first m eigenvalues (counting multiplicities) in $\sigma(L_\lambda)$:

$$\mathrm{card}(\sigma_{\mathfrak{c}}(L_\lambda)) = m. \tag{3.13}$$

It is interesting to note that the uniform spectrum decomposition (3.11) prevents eigenvalues in $\sigma_{\mathfrak{s}}(L_\lambda)$ from merging with eigenvalues in $\sigma_{\mathfrak{c}}(L_\lambda)$ as λ varies in Λ, while the cardinality of $\sigma_{\mathfrak{c}}(L_\lambda)$ remains fixed to be m over Λ. As a consequence, the spaces $\mathcal{H}_\alpha$ and $\mathcal{H}$ can be decomposed into L_λ-invariant subspaces in such a way that the subspace associated with $\sigma_{\mathfrak{c}}(L_\lambda)$, $\mathcal{H}^{\mathfrak{c}}(\lambda)$ defined below in (3.18), has fixed dimension m for each $\lambda \in \Lambda$. These subspaces will be at the basis of the construction of stochastic invariant manifolds considered in later chapters. Furthermore, the uniform spectrum decomposition (3.11) will be essential to guarantee the existence of a neighborhood of the origin (in $\mathcal{H}^{\mathfrak{c}}(\lambda)$) whose diameter is independent of $\lambda \in \Lambda$ over which a family of stochastic local invariant manifolds—each of dimension m—is defined; see Theorem 5.1 and Corollary 5.1. The existence of such a neighborhood will turn out to be particularly useful in the examination of stochastic bifurcations or more general phase transitions associated with SPDEs of type (3.1); see Remark 6.3 and [36].

Remark 3.1 Note that $\eta_{\mathfrak{c}}$ and $\eta_{\mathfrak{s}}$ are allowed to share the same sign in the derivation of all the results of Chaps. 4 and 5. The results of Chap. 6 are presented only in the case where $\eta_{\mathfrak{s}} < \eta_{\mathfrak{c}} < 0$, which will be sufficient for the applications dealt with in [36].

Related L_λ-invariant subspaces. We present now decompositions of the spaces $\mathcal{H}$ and $\mathcal{H}_\alpha$ into L_λ-invariant subspaces naturally associated with the splitting of the spectrum $\sigma(L_\lambda)$ given in (3.11). These decompositions lead naturally to a partial dichotomy of the deterministic linear semigroup[5] associated with (3.1); see (3.24a)–(3.24c).

This partial dichotomy will turn out to be sufficient in the construction of stochastic invariant manifolds for the full SPDE, as we will see in the forthcoming chapters. The reason is that only stochastic invariant manifolds associated with the trivial steady state are considered in this monograph. For manifolds associated with more general (random) stationary solutions, other types of dichotomy estimates which involve the Lyapunov spectrum are typically required; see Remark 3.2. To simplify the presentation of our main results, such manifolds will not be considered in this monograph.

The decompositions of $\mathcal{H}$ and $\mathcal{H}_\alpha$ are organized as follows. Let us first introduce the complexifications, $\widetilde{\mathcal{H}_1}$ and $\widetilde{\mathcal{H}}$, of the spaces $\mathcal{H}_1$ and $\mathcal{H}$:

$$\widetilde{\mathcal{H}_1} := \{u + iv \mid u, v \in \mathcal{H}_1\}, \quad \widetilde{\mathcal{H}} := \{u + iv \mid u, v \in \mathcal{H}\}, \tag{3.14}$$

where i denotes here the imaginary unit. Denote also the complexification of L_λ by $\widetilde{L}_\lambda$, i.e.,

$$\widetilde{L}_\lambda(u + iv) := L_\lambda u + iL_\lambda v, \quad \forall\, u, v \in \mathcal{H}_1. \tag{3.15}$$

[5] Namely, the semigroup associated with $dv = L_\lambda v\, dt$.

Note that for each $\lambda \in \Lambda$, the set $\sigma_{\mathfrak{c}}(L_\lambda)$ in the uniform spectrum decomposition (3.11) is bounded. Hence, according to [104, Theorem III-6.17] (see also [71, Proposition IV.1.16]), there exists a projector $\widetilde{P}_{\mathfrak{c}}$ associated with $\sigma_{\mathfrak{c}}(L_\lambda)$, such that the space $\widetilde{\mathscr{H}}$ can be decomposed as follows:

$$\widetilde{\mathscr{H}} = \widetilde{\mathscr{H}^{\mathfrak{c}}}(\lambda) \oplus \widetilde{\mathscr{H}^{\mathfrak{s}}}(\lambda), \tag{3.16}$$

where

$$\widetilde{\mathscr{H}^{\mathfrak{c}}}(\lambda) := \widetilde{P}_{\mathfrak{c}}\widetilde{\mathscr{H}}, \quad \widetilde{\mathscr{H}^{\mathfrak{s}}}(\lambda) := \big(\mathrm{Id}_{\widetilde{\mathscr{H}}} - \widetilde{P}_{\mathfrak{c}}\big)\widetilde{\mathscr{H}}. \tag{3.17}$$

Note that $\widetilde{P}_{\mathfrak{c}}$ is given as the Riesz projector defined by

$$\widetilde{P}_{\mathfrak{c}} := -\frac{1}{2\pi i}\int_{\Gamma_{\mathfrak{c}}} (\widetilde{L}_\lambda - \beta \mathrm{Id})^{-1} \mathrm{d}\beta,$$

where $\Gamma_{\mathfrak{c}}$ is a rectifiable closed curve surrounding the eigenvalues of $\sigma_{\mathfrak{c}}(L_\lambda)$ which does not include any elements of $\sigma_{\mathfrak{s}}(L_\lambda)$.

Moreover, $\widetilde{\mathscr{H}^{\mathfrak{c}}}(\lambda)$ and $\widetilde{\mathscr{H}^{\mathfrak{s}}}(\lambda)$ are invariant under $\widetilde{L}_\lambda$ in the following sense [146, Theorem 5.7-A]:

$$\widetilde{L}_\lambda(\widetilde{\mathscr{H}^{\mathfrak{c}}}(\lambda) \cap \widetilde{\mathscr{H}_1}) \subset \widetilde{\mathscr{H}^{\mathfrak{c}}}(\lambda), \quad \widetilde{L}_\lambda(\widetilde{\mathscr{H}^{\mathfrak{s}}}(\lambda) \cap \widetilde{\mathscr{H}_1}) \subset \widetilde{\mathscr{H}^{\mathfrak{s}}}(\lambda);$$

and the restriction of $\widetilde{L}_\lambda$ on $\widetilde{\mathscr{H}^{\mathfrak{c}}}(\lambda)$, denoted by $\widetilde{L}_\lambda^{\mathfrak{c}}$, is a bounded linear operator on $\widetilde{\mathscr{H}^{\mathfrak{c}}}(\lambda)$ according to [146, Theorem 5.8-A] or [104, Theorem III-6.17].

Now, for $\lambda \in \Lambda$, let us define

$$\mathscr{H}^{\mathfrak{c}}(\lambda) := \{u, v \mid u, v \in \mathscr{H}, \text{ and } u + iv \in \widetilde{\mathscr{H}^{\mathfrak{c}}}(\lambda)\}. \tag{3.18}$$

Note that $\mathscr{H}^{\mathfrak{c}}(\lambda)$ thus defined forms naturally a subspace of $\mathscr{H}$; and it can be checked that[6]

$$\dim \mathscr{H}^{\mathfrak{c}}(\lambda) = \dim_{\mathbb{C}} \widetilde{\mathscr{H}^{\mathfrak{c}}}(\lambda) = m, \quad \forall\, \lambda \in \Lambda, \tag{3.19}$$

where m is the cardinality of $\sigma_c(L_\lambda)$ as given in (3.13).

---

[6]Equation (3.19) can be justified as follows. Since L_λ has real coefficients, for any complex eigenvalue $\beta_j(\lambda) \in \sigma_{\mathfrak{c}}(L_\lambda)$, its conjugate $\overline{\beta}_j(\lambda)$ is also an eigenvalue which belongs to $\sigma_{\mathfrak{c}}(L_\lambda)$.

After possibly reordering the eigenvalues, the space $\widetilde{\mathscr{H}^{\mathfrak{c}}}(\lambda)$ can be further decomposed into $\bigoplus_{j=1}^{l}\big(\widetilde{\mathscr{H}_{\beta_j(\lambda)}} \oplus \widetilde{\mathscr{H}_{\overline{\beta}_j(\lambda)}}\big) \bigoplus_{j=2l+1}^{m} \widetilde{\mathscr{H}_{\beta_j(\lambda)}}$, where $\widetilde{\mathscr{H}_{\beta_j(\lambda)}}$ is the eigenspace associated with $\beta_j(\lambda)$, and the first $2l$ eigenvalues are (genuinely) complex and the remaining are real.

For each $\widetilde{\mathscr{H}_{\beta_j(\lambda)}}$, a real vector space $\mathscr{H}_{\beta_j(\lambda)}$ can be defined in the same way as in (3.18). Note that $\mathscr{H}_{\beta_j(\lambda)} = \mathscr{H}_{\overline{\beta}_j(\lambda)}$, for all $j \in \{1, \ldots, l\}$. Then, $\mathscr{H}^{\mathfrak{c}}(\lambda)$ admits the following decomposition: $\bigoplus_{j=1}^{l} \mathscr{H}_{\beta_j(\lambda)} \bigoplus_{j=2l+1}^{m} \mathscr{H}_{\beta_j(\lambda)}$. Since $\dim \mathscr{H}_{\beta_j(\lambda)} = \dim_{\mathbb{C}} \widetilde{\mathscr{H}_{\beta_j(\lambda)}} + \dim_{\mathbb{C}} \widetilde{\mathscr{H}_{\overline{\beta}_j(\lambda)}}$ for all $j \in \{1, \ldots, l\}$, and $\dim \mathscr{H}_{\beta_j(\lambda)} = \dim_{\mathbb{C}} \widetilde{\mathscr{H}_{\beta_j(\lambda)}}$ for all $j \in \{2l+1, \ldots, m\}$, the result follows.

The space $\mathscr{H}^{\mathfrak{c}}(\lambda)$ uniquely determines the closed subspaces $\mathscr{H}^{\mathfrak{s}}(\lambda)$ and $\mathscr{H}_\alpha^{\mathfrak{s}}(\lambda)$ as the topological complements in $\mathscr{H}$ and $\mathscr{H}_\alpha$ respectively, i.e.,

$$\mathscr{H} = \mathscr{H}^{\mathfrak{c}}(\lambda) \oplus \mathscr{H}^{\mathfrak{s}}(\lambda), \quad \mathscr{H}_\alpha = \mathscr{H}^{\mathfrak{c}}(\lambda) \oplus \mathscr{H}_\alpha^{\mathfrak{s}}(\lambda), \quad \forall\, \lambda \in \Lambda. \tag{3.20}$$

Let

$$P_{\mathfrak{c}}(\lambda) : \mathscr{H} \to \mathscr{H}^{\mathfrak{c}}(\lambda), \quad P_{\mathfrak{s}}(\lambda) : \mathscr{H} \to \mathscr{H}^{\mathfrak{s}}(\lambda) \tag{3.21}$$

be the associated canonical (spectral) projectors, and we denote

$$L_\lambda^{\mathfrak{c}} := L_\lambda P_{\mathfrak{c}}(\lambda), \quad L_\lambda^{\mathfrak{s}} := L_\lambda P_{\mathfrak{s}}(\lambda). \tag{3.22}$$

Note that L_λ commutes with $P_{\mathfrak{c}}(\lambda)$ and $P_{\mathfrak{s}}(\lambda)$, which follows from the fact that $\widetilde{\mathscr{H}}^{\mathfrak{c}}(\lambda)$ and $\widetilde{\mathscr{H}}^{\mathfrak{s}}(\lambda)$ are invariant under $\widetilde{L}_\lambda$. As a consequence, the subspaces $\mathscr{H}^{\mathfrak{c}}(\lambda)$ and $\mathscr{H}^{\mathfrak{s}}(\lambda)$ are invariant by the semigroup e^{tL_λ}. Note also that similar to $\widetilde{L}_\lambda^{\mathfrak{c}}$, the operator $L_\lambda^{\mathfrak{c}}$ is a bounded linear operator on $\mathscr{H}^{\mathfrak{c}}(\lambda)$, so that $e^{tL_\lambda} P_{\mathfrak{c}}$ can be extended to $t < 0$, namely $e^{tL_\lambda} P_{\mathfrak{c}}$ defines a flow on $\mathscr{H}^{\mathfrak{c}}(\lambda)$. This fact is used in the partial dichotomy estimate (3.24c) below.

As mentioned above, note that by (3.19), the dimension of $\mathscr{H}^{\mathfrak{c}}(\lambda)$ is independent of λ as it varies in Λ, so that $\mathscr{H}^{\mathfrak{c}}(\lambda)$ is unique up to orthogonal transformations. For the sake of concision, this property has led us to suppress the λ-dependence of the subspaces given in (3.20), and of the projectors $P_{\mathfrak{c}}(\lambda)$ and $P_{\mathfrak{s}}(\lambda)$ defined in (3.21). The results are derived and presented hereafter according to this convention.

Partial-dichotomy estimates. Thanks to the uniform spectrum decomposition (3.11), for any given numbers η_1 and η_2 satisfying

$$\eta_{\mathfrak{c}} > \eta_1 > \eta_2 > \eta_{\mathfrak{s}}, \tag{3.23}$$

there exists a constant, $K \geq 1$, such that for all $\lambda \in \Lambda$ the following *partial-dichotomy*[7] estimates hold for the semigroup generated by L_λ (see, e.g., [92, Theorems 1.5.3, 1.5.4][8]):

$$\|e^{tL_\lambda} P_{\mathfrak{s}}\|_{L(\mathscr{H}_\alpha, \mathscr{H}_\alpha)} \leq K e^{\eta_2 t}, \quad t \geq 0, \tag{3.24a}$$

$$\|e^{tL_\lambda} P_{\mathfrak{s}}\|_{L(\mathscr{H}, \mathscr{H}_\alpha)} \leq K t^{-\alpha} e^{\eta_2 t}, \quad t > 0, \tag{3.24b}$$

$$\|e^{tL_\lambda} P_{\mathfrak{c}}\|_{L(\mathscr{H}, \mathscr{H}_\alpha)} \leq K e^{\eta_1 t}, \quad t \leq 0, \tag{3.24c}$$

[7] The partial aspect of the dichotomy is explained when η_1 and η_2 share the same sign which is allowed by (3.11); see also Remark 3.1. In that case, the distinction is made on the magnitude of the rate of contraction (or expansion) associated with $\mathrm{d}v = L_\lambda v \mathrm{d}t$; otherwise the concept matches the classical one of exponential dichotomy found in the literature; see, e.g., [92, 143].

[8] The first two inequalities in (3.24a)–(3.24c) follow from [92, Theorem 1.5.4] by choosing the operator A thereof to be $-L_\lambda$ and the operator B to be our operator A; the third inequality follows from [92, Theorem 1.5.3], the L_λ-invariance of $\mathscr{H}^{\mathfrak{c}}$, and the fact that any norms on $\mathscr{H}^{\mathfrak{c}}$ are equivalent since it is finite dimensional.

where $L(X, Y)$ denotes the space of bounded linear operators from the Banach space X to the Banach space Y. Note that the estimate given in (3.24b) accounts for the instantaneous smoothing effects of the semigroup e^{tL_λ} for $t > 0$ from $\mathscr{H}$ to $\mathscr{H}_\alpha$ where we recall that $\mathscr{H}_\alpha$ has been imposed by the choice of the nonlinearity.

The conditions $\eta_\mathfrak{c} > \eta_1$ and $\eta_2 > \eta_\mathfrak{s}$ allow us to absorb the polynomial growth terms in the estimates (3.24a)–(3.24c) that—because of our assumptions (L_λ being not necessarily self-adjoint)—could be present in front of the exponential terms with $\eta_\mathfrak{c}$ (resp. $\eta_\mathfrak{s}$) in place of η_1 (resp. η_2). As a consequence, K in (3.24a)–(3.24c) depends on $\eta_* := \min\{\eta_\mathfrak{c} - \eta_1, \eta_2 - \eta_\mathfrak{s}\}$, and may get larger as η_* gets closer to zero in the non-self-adjoint case. Note that however, K is independent of $\lambda \in \Lambda$ in all the cases.

Remark 3.2

(1) It is important to note that in this monograph, for the sake of simplicity, the stochastic critical manifold theory along with the corresponding approximation formulas derived in Chap. 6, are presented in the vicinity of the trivial steady state, as λ varies in a neighborhood of the critical parameter λ_c. In Volume II [37], we will consider manifolds which are not subject to such a locality requirement.
(2) Throughout this monograph, the use of a random frame which moves with the cocycle, will not be required to derive the approximation formulas concerned with Theorem 6.1. As a consequence, the approach presented here does not make usage of the Lyapunov spectrum and the multiplicative ergodic theory (MET) in Hilbert or Banach spaces [113, 141]. The MET is typically employed when stochastic invariant manifolds in the vicinity of a nontrivial random stationary solution are concerned; see, e.g., [27, 113, 124]. We leave for future research the extension of the approximation formulas derived in this monograph to a MET setting inclined to deal with more general noises; see however [35] for an alternative approach in the case of additive noise.

3.2 Random Dynamical Systems

In this section, we recall the definitions of metric dynamical systems (MDSs) and random dynamical systems (RDSs), and specify—in a measure-theoretic sense—the canonical MDS associated with the Wiener process in Eq. (3.1) which will be used throughout this monograph. The interested readers are referred to [1, 47, 54] for more details, and to [39] for an intuitive and "physically-oriented" presentation of these concepts.

Metric dynamical system. A family of mappings $\{\theta_t\}_{t\in\mathbb{R}}$ on a probability space $(\Omega, \mathscr{F}, \mathbb{P})$ is called a *metric dynamical system* if the following conditions are satisfied:

(i) $(t, \omega) \mapsto \theta_t\omega$ is $(\mathscr{B}(\mathbb{R}) \otimes \mathscr{F}; \mathscr{F})$-measurable, where $\mathscr{B}(\mathbb{R})$ denotes the Borel σ-algebra on $\mathbb{R}$, and $\mathscr{B}(\mathbb{R}) \otimes \mathscr{F}$ denotes the σ-algebra generated by the direct product of elements of $\mathscr{B}(\mathbb{R})$ and $\mathscr{F}$;

(ii) $\{\theta_t\}$ satisfies the one-parameter group property, i.e., $\theta_0 = \mathrm{Id}_\Omega$, and $\theta_{t+s} = \theta_t \circ \theta_s$ for all $t, s \in \mathbb{R}$;

(iii) $\mathbb{P}$ is invariant with respect to θ_t for all $t \in \mathbb{R}$, i.e., $(\theta_t)_*\mathbb{P} = \mathbb{P}$ for all $t \in \mathbb{R}$, where $(\theta_t)_*\mathbb{P}$ is the *push-forward measure* of $\mathbb{P}$ by θ_t, defined by $(\theta_t)_*(F) := \mathbb{P}(\theta_{-t}(F))$, for all $F \in \mathscr{F}$.

Continuous random dynamical system. Given a separable Hilbert space $(H, |\cdot|_H)$ with the associated Borel σ-algebra denoted by $\mathscr{B}(H)$, a continuous *random dynamical system* acting on H over an MDS, $(\Omega, \mathscr{F}, \mathbb{P}, \{\theta_t\}_{t\in\mathbb{R}})$, is a $(\mathscr{B}(\mathbb{R}^+) \otimes \mathscr{F} \otimes \mathscr{B}(H); \mathscr{B}(H))$-measurable mapping

$$S\colon \mathbb{R}^+ \times \Omega \times H \to H, \quad (t, \omega, \xi) \mapsto S(t, \omega)\xi,$$

which satisfies the following properties:

(i)′ $S(0, \omega) = \mathrm{Id}_H$, for all $\omega \in \Omega$,

(ii)′ S satisfies the *perfect cocycle property*, i.e.,

$$S(t+s, \omega) = S(t, \theta_s\omega) \circ S(s, \omega), \quad \forall\ t, s \in \mathbb{R}^+, \text{ and } \omega \in \Omega,$$

(iii)′ $S(t, \omega)\colon H \to H$ is continuous for all $t \in \mathbb{R}^+$ and $\omega \in \Omega$.

We are in position to introduce the MDS that we will work with throughout this monograph. Let us first recall the canonical MDS, $(\Omega, \mathscr{F}, \mathbb{P}, \{\theta_t\}_{t\in\mathbb{R}})$, associated with the Wiener process; see, e.g., [1, Appendices A.2 and A.3] and [47, Chap. 1]. Here the sample space Ω consists of the sample paths of a two-sided one-dimensional Wiener process W_t taking zero value at $t = 0$, that is,

$$\Omega = \{\omega \in C(\mathbb{R}, \mathbb{R}) \mid \omega(0) = 0\};$$

$\mathscr{F}$ is the Borel σ-algebra associated with the Wiener process; $\mathbb{P}$ is the classical Wiener measure on Ω; and for each $t \in \mathbb{R}$, the map $\theta_t\colon (\Omega, \mathscr{F}, \mathbb{P}) \to (\Omega, \mathscr{F}, \mathbb{P})$ is the measure preserving transformation defined by:

$$\theta_t\omega(\cdot) = \omega(\cdot + t) - \omega(t). \tag{3.25}$$

In order the solution operator associated with Eq. (3.1) to satisfy the perfect cocycle property given in (ii)′ above, we will restrict our consideration to some subset in Ω of full measure which is also θ_t-invariant for all $t \in \mathbb{R}$. In the following, we will identify such a subset, and introduce the restriction of the canonical MDS to this subset.

In that respect, let us consider the following scalar Langevin equation:

$$\mathrm{d}z + z\,\mathrm{d}t = \sigma\,\mathrm{d}W_t. \tag{3.26}$$

It is known that this equation possesses a unique stationary solution $z_\sigma(\theta_t\omega)$—the stationary Ornstein-Uhlenbeck (OU) process—whose main properties are in particular recalled in the following lemma.

Lemma 3.1 *There exists a subset Ω^* of Ω which is of full measure and is θ_t-invariant for all $t \in \mathbb{R}$, i.e.,*

$$\mathbb{P}(\Omega^*) = 1, \quad \text{and} \quad \theta_t(\Omega^*) = \Omega^* \quad \forall\, t \in \mathbb{R}; \tag{3.27}$$

and the following properties hold on Ω^*:

(i) *For each $\omega \in \Omega^*$, $t \mapsto W_t(\omega)$ is γ-Hölder for any $\gamma \in (0, 1/2)$.*
(ii) *$t \mapsto W_t(\omega)$ has sublinear growth:*

$$\lim_{t\to\pm\infty} \frac{W_t(\omega)}{t} = 0, \quad \forall\, \omega \in \Omega^*. \tag{3.28}$$

(iii) *For each $\omega \in \Omega^*$, $t \mapsto z_\sigma(\theta_t\omega)$ is γ-Hölder for any $\gamma \in (0, 1/2)$, and can be written as:*

$$\begin{aligned} z_\sigma(\theta_t\omega) &= -\sigma \int_{-\infty}^{0} e^\tau W_\tau(\theta_t\omega)\, d\tau \\ &= -\sigma \int_{-\infty}^{0} e^\tau W_{\tau+t}(\omega)\, d\tau + \sigma W_t(\omega), \quad t \in \mathbb{R}, \quad \omega \in \Omega^*. \end{aligned} \tag{3.29}$$

(iv) *The following growth control relations are satisfied:*

$$\lim_{t\to\pm\infty} \frac{z_\sigma(\theta_t\omega)}{t} = 0, \ \text{and} \ \lim_{t\to\pm\infty} \frac{1}{t}\int_0^t z_\sigma(\theta_\tau\omega)\, d\tau = 0, \quad \forall\, \omega \in \Omega^*. \tag{3.30}$$

Proof It is known that there exists a subset $\Omega' \subset \Omega$ of full measure which is θ_t-invariant for all $t \in \mathbb{R}$, and (ii), (iv) and (3.29) hold on Ω'; see, e.g., [28, 47, 65].[9]

For the sake of clarity, we explain here how to exhibit a subset $\Omega'' \subset \Omega$ of full measure which is θ_t-invariant and for which $t \mapsto z_\sigma(\theta_t\omega)$ is (locally) γ-Hölder for any $\gamma \in (0, 1/2)$ and any $\omega \in \Omega''$. Note that by simply integrating Eq. (3.26) we get trivially that:

[9]Regarding (3.30) see also [28, Lemma 4.1], and the Birkhoff-Khinchin ergodic theorem in e.g. [1, p. 539]. We will make very often use of these growth control relations in the proofs of Theorems 4.3 and 6.1 presented below.

$$\int_0^t z_\sigma(\theta_s\omega)\,\mathrm{d}s + z_\sigma(\theta_t\omega) = z_\sigma(\omega) + \sigma W_t(\omega), \qquad \forall\, t \in \mathbb{R}. \tag{3.31}$$

From this identity, it is thus sufficient to exhibit such an Ω'' where the γ-Hölder property holds for $t \mapsto W_t$.

It is known that for each $m \in \mathbb{N}^*$ the Wiener process is almost surely γ-Hölder for any exponent $\gamma < \frac{1}{2} - \frac{1}{2m}$; see, e.g., [73, Sect. 3.4]. For each $m \geq 1$, let Ω_m be a corresponding subset of full measure. By possibly redefining Ω_m to be $\Omega_m \cap \Omega_{m-1}$ so that $\Omega_m \subset \Omega_{m-1}$, we deduce naturally that

$$\Omega'' := \bigcap_{m=1}^{\infty} \Omega_m \tag{3.32}$$

is such that $\mathbb{P}(\Omega'') = \lim_{m\to\infty} \mathbb{P}(\Omega_m) = 1$; see, e.g., [82, Theorem 1.8].

By construction of Ω'', any sample path ω in Ω'' is γ-Hölder for any exponent less than $1/2$. Trivially, $\theta_t\omega$ shares this Hölder property for all $\omega \in \Omega''$ since θ_t is just a shift on Ω; see (3.25). The set Ω'' is thus θ_t-invariant for all $t \in \mathbb{R}$, and the proof is complete by taking $\Omega^* := \Omega' \cap \Omega''$. □

Now, let Ω^* be the θ_t-invariant subset of Ω as given in Lemma 3.1, and $\mathcal{F}_{\Omega^*}$ be the trace σ-algebra of $\mathcal{F}$ with respect to Ω^*, i.e.,

$$\mathcal{F}_{\Omega^*} := \{F \cap \Omega^* \mid F \in \mathcal{F}\}.$$

It can be checked that $(t,\omega) \to \theta_t\omega$ is $(\mathcal{B}(\mathbb{R}) \otimes \mathcal{F}_{\Omega^*}; \mathcal{F}_{\Omega^*})$-measurable; see for instance [27, Lemma 3.2]. It follows that $(\Omega^*, \mathcal{F}_{\Omega^*}, \mathbb{P}_{\Omega^*}, \{\theta_t\}_{t\in\mathbb{R}})$ forms an MDS, where $\mathbb{P}_{\Omega^*}$ denotes the restriction of $\mathbb{P}$ on Ω^*.

To simplify the notations and the presentation, we will denote hereafter the new sample space $(\Omega^*, \mathcal{F}_{\Omega^*}, \mathbb{P}_{\Omega^*})$ as $(\Omega, \mathcal{F}, \mathbb{P})$; and we will work with this restricted MDS, $(\Omega, \mathcal{F}, \mathbb{P}, \{\theta_t\}_{t\in\mathbb{R}})$, without confusion with the original MDS. For the sake of concision, we will often omit mentioning such an underlying MDS, thus identifying an RDS with its cocycle part.

3.3 Cohomologous Cocycles and Random Evolution Equations

Our approach to develop a theory of stochastic critical manifolds for SPDEs of type (3.1) along with the corresponding approximation formulas, will be rooted in the use of a smooth cohomology,[10] i.e., a stationary coordinate change, on the

[10] The theory of stochastic parameterizing manifolds presented in Volume II [37] will also make use of such a cohomology.

phase space $\mathscr{H}_\alpha$. Such a strategy was initially motivated by the study of certain asymptotic problems arising with finite-dimensional SDEs; see [1, 95–97]. The idea is to transform the original SDE into a random differential equation (RDE), i.e., a randomly parameterized ordinary differential equation, which is *de facto* more amenable for the analysis of random attractors or random invariant manifolds since these objects require a pathwise framework from their definitions. The cohomology offers then a sort of *transfer principle for these asymptotic problems*: if a result is proved on, say, random invariant manifolds for the transformed RDE, then it holds true for the original SDE up to a stationary conjugacy. This approach was adopted and extended to the context of SPDEs driven by multiplicative noise in [77] for the study of random attractors, and in [66] for the study of stochastic invariant manifolds.

We follow here more specifically the approach of [66] to set up the original SPDE (3.1) within the RDS framework. In that respect, we transform Eq. (3.1) into an evolution equation with random coefficients which helps simplify the analysis of the dynamics associated with the SPDE.

As mentioned above, such a simplification is rooted in the interpretation—up to a smooth cohomology—of the solutions of Eq. (3.1) as an ω-wise version of classical solutions of the transformed equation whose existence can be guaranteed from the literature of non-autonomous deterministic evolution equations [92, 138, 143, 147] although the measurability properties of such solutions require a special attention, which is carried out in Appendix A.[11]

Let us now introduce the following standard change of variables:

$$v(t) = e^{-z_\sigma(\theta_t\omega)}u(t), \tag{3.33}$$

where z_σ is the OU process defined in (3.29).

Note that by the Itô formula (cf. [130, Theorem 4.1.2]), the stochastic process $e^{-z_\sigma(\theta_t\omega)}$ satisfies

$$\begin{aligned} \mathrm{d}e^{-z_\sigma(\theta_t\omega)} &= \Big(z_\sigma(\theta_t\omega)e^{-z_\sigma(\theta_t\omega)} + \frac{\sigma^2}{2}e^{-z_\sigma(\theta_t\omega)}\Big)\mathrm{d}t - \sigma e^{-z_\sigma(\theta_t\omega)}\mathrm{d}W_t \\ &= z_\sigma(\theta_t\omega)e^{-z_\sigma(\theta_t\omega)}\mathrm{d}t - \sigma e^{-z_\sigma(\theta_t\omega)} \circ \mathrm{d}W_t, \end{aligned} \tag{3.34}$$

where the second equality above follows from the conversion between the Itô and Stratonovich integrals; cf. [110, Theorem 2.3.11].

Formally, we also have that

$$\mathrm{d}v = \mathrm{d}\big(e^{-z_\sigma(\theta_t\omega)}u\big) = u \circ \mathrm{d}e^{-z_\sigma(\theta_t\omega)} + e^{-z_\sigma(\theta_t\omega)} \circ \mathrm{d}u. \tag{3.35}$$

---

[11] Note that a direct approach at the level of the SPDE (i.e., without using random transformations) could have been used to put Eq. (3.1) within the RDS framework [124]. We adopted here the cohomology approach [95–97] which makes possible the use of standard estimation techniques that do not rely on stochastic analysis [58]. Of course the simple nature of our multiplicative noise makes efficient such a strategy here. For more general multiplicative noise, techniques from [77] could be combined with the ones presented hereafter, to extend the main results of Chaps. 6 and 7.

Then, by using Eqs. (3.1) and (3.34) into the above equation, we find after simplification that v satisfies formally the following random evolution equation (hereafter referred to as an RPDE):

$$\frac{\mathrm{d}v}{\mathrm{d}t} = L_\lambda v + z_\sigma(\theta_t\omega)v + G(\theta_t\omega,\, v), \tag{3.36}$$

where $G(\omega,\, v) := e^{-z_\sigma(\omega)}F(e^{z_\sigma(\omega)}v)$. Note that $G(\omega, v)$ is globally Lipschitz in v and has the same Lipschitz constant as F; and $G(\omega,\, 0) = 0$ for all $\omega \in \Omega$.

As mentioned above, the existence of pathwise solutions to Eq. (3.36) for each fixed ω can be guaranteed from the literature of non-autonomous deterministic evolution equations since in particular the OU process z_σ is Hölder continuous in time; see Lemma 3.1. Only the measurability properties of such solutions need a particular attention. We summarize the precise results in the following proposition. A full proof is provided in Appendix A for the reader's convenience.

Proposition 3.1 *Consider the RPDE* (3.36). *The assumptions on L_λ and F are those of Sect.* 3.1 *where F is assumed to be globally Lipschitz here; see* (3.7).

Then, for each $v_0 \in \mathcal{H}_\alpha$, and $\omega \in \Omega$, Eq. (3.36) *has a unique classical solution $v_\lambda(t, \omega; v_0)$ with initial datum $v_\lambda(0, \omega; v_0) = v_0$ such that*

$$v_\lambda(\cdot, \omega; v_0) \in C((0, T]; \mathcal{H}_1) \cap C([0, T]; \mathcal{H}_\alpha) \cap C^1((0, T]; \mathcal{H}), \quad \forall T > 0. \tag{3.37}$$

Furthermore, for each λ, v_λ is $\big(\mathcal{B}(\mathbb{R}^+) \otimes \mathcal{F} \otimes \mathcal{B}(\mathcal{H}_\alpha); \mathcal{B}(\mathcal{H}_\alpha)\big)$-measurable; and for each ω, $v_\lambda(\cdot, \omega; v_0)$ depends continuously on λ and v_0.

It is worthwhile to remark that the conditions under which the conclusions of Proposition 3.1 hold are obviously not optimal but sufficient for our purpose. For instance, the Lipschitz condition on the nonlinearity can be replaced by certain dissipative conditions on the nonlinear terms which hold for a broad class of physical problems. We refer again to the aforementioned works [33, 92, 138, 143, 147] for more details; see also [37, 58, 76, 85].

A direct consequence of this proposition is that, for any λ and any $(\mathcal{F}; \mathcal{B}(\mathcal{H}_\alpha))$-measurable random initial datum $v_0(\omega)$, there exists a unique classical solution $v_{\lambda, v_0(\omega)}(t, \omega) := v_\lambda(t, \omega; v_0(\omega))$ of Eq. (3.36) which is $\big(\mathcal{B}(\mathbb{R}^+) \otimes \mathcal{F}; \mathcal{B}(\mathcal{H}_\alpha)\big)$-measurable, and $v_{\lambda, v_0(\omega)}(\cdot, \omega)$ has the regularity specified in (3.37) for each ω.

Based on this proposition, we can define for each λ, an RDS generated by Eq. (3.36), S_λ, as follows:

$$S_\lambda : \mathbb{R}^+ \times \Omega \times \mathcal{H}_\alpha \to \mathcal{H}_\alpha, \quad (t, \omega, v_0) \mapsto S_\lambda(t, \omega)v_0 := v_\lambda(t, \omega; v_0), \tag{3.38}$$

where $v_\lambda(t, \omega; v_0)$ is the global classical solution to Eq. (3.36) with initial datum v_0, which can be either deterministic or an $\mathcal{H}_\alpha$-valued random variable.

We now introduce the random (smooth) transformation acting on the space $\mathcal{H}_\alpha$,

$$\mathfrak{D}(\omega)\xi := \xi e^{-z_\sigma(\omega)}, \tag{3.39}$$

and its inverse random (smooth) transformation,

$$\mathfrak{D}^{-1}(\omega)\xi := \xi e^{z_\sigma(\omega)}, \tag{3.40}$$

where $\xi \in \mathscr{H}_\alpha$, and $\omega \in \Omega$.

Let us now define the mapping $\widehat{S}_\lambda : \mathbb{R}^+ \times \Omega \times \mathscr{H}_\alpha \to \mathscr{H}_\alpha$ via

$$\widehat{S}_\lambda(t, \omega)u_0 := \mathfrak{D}^{-1}(\theta_t\omega) \circ S_\lambda(t, \omega) \circ \mathfrak{D}(\omega)u_0, \tag{3.41}$$

where the symbol, $\circ$, denotes the basic composition operation between two self-mappings on $\mathscr{H}_\alpha$. The mapping $\widehat{S}_\lambda$ thus defined is clearly measurable, and defines an RDS acting on $\mathscr{H}_\alpha$.

By a solution to the SPDE (3.1) with initial datum $u_0 \in \mathscr{H}_\alpha$, we always mean a process $u_\lambda(t, \omega; u_0) := e^{z_\sigma(\theta_t\omega)} v_\lambda(t, \omega; u_0 e^{-z_\sigma(\omega)})$, where v_λ is a classical solution of the RPDE (3.36) with initial datum $u_0 e^{-z_\sigma(\omega)}$. In that sense, the RDS, $\widehat{S}_\lambda$, provides solutions to Eq. (3.1) since $u_\lambda(t, \omega; u_0) = \widehat{S}_\lambda(t, \omega)u_0$.

Remark 3.3 It is important to note in (3.41) that, while the cocycles $S_\lambda(t, \omega)$ and $\widehat{S}_\lambda(t, \omega)$ map the ω-fiber to the $\theta_t\omega$-fiber, the bijective random coordinate transformation $\mathfrak{D}$ maps only each fiber to itself, explaining the shift of fiber appearing in the inverse transformation $\mathfrak{D}^{-1}$.

The two random dynamical systems S_λ and $\widehat{S}_\lambda$ are thus *cohomologous* [1, 95] via the random C^∞-diffeomorphism $\mathfrak{D}$ acting on $\mathscr{H}_\alpha$. As pointed out above, the phase portrait of the dynamics associated with Eq. (3.1), in $\mathscr{H}_\alpha$, is thus fiber-wise C^∞-diffeomorphic to the phase portrait of the dynamics associated with Eq. (3.36). Any regularity as well as topological properties of an invariant manifold for S_λ is therefore preserved for $\widehat{S}_\lambda$. Since the random evolution equation (3.36) is in a more convenient form for the analysis developed hereafter, we will keep proving our results for this equation. The corresponding results associated with Eq. (3.1) will be then transferred naturally from those proved for (3.36) by application of the cohomology relation (3.41); see, e.g., Corollaries 4.1–4.3, and 5.1.

3.4 Linearized Stochastic Flow and Related Estimates

The solution operator $\mathfrak{T}_{\lambda,\sigma}$ associated with the linearized equation (about the trivial steady state),

$$\frac{\mathrm{d}v}{\mathrm{d}t} = L_\lambda v + z_\sigma(\theta_t\omega)v, \tag{3.42}$$

will play a key role in most of the estimates carried out in the forthcoming chapters arising in various fixed point problems involving nonlinear integral equations associated with Eq. (3.1). We provide here in that respect some related estimates on $\mathfrak{T}_{\lambda,\sigma}$.

First note that since $z_\sigma(\theta_t\omega)\mathrm{Id}$ commutes obviously with $z_\sigma(\theta_{t'}\omega)\mathrm{Id}$ when $t \neq t'$, we obtain thus that $e^{\int_0^t z_\sigma(\theta_\tau\omega)\,\mathrm{d}\tau \mathrm{Id}}$ defines the solution operator of the equation $\frac{\mathrm{d}v}{\mathrm{d}t} = z_\sigma(\theta_t\omega)v$. Now, for any given $t \geq 0$, by using the basic random change of variables $\widetilde{v} := e^{-\int_0^t z_\sigma(\theta_\tau\omega)\,\mathrm{d}\tau \mathrm{Id}}v$, it can be checked that $v(t,\omega;v_0)$ solves Eq. (3.42) if and only if $\widetilde{v}(t;v_0)$ solves

$$\frac{\mathrm{d}\widetilde{v}}{\mathrm{d}t} = L_\lambda \widetilde{v}, \tag{3.43}$$

where $\mathrm{Id}\colon \mathscr{H} \to \mathscr{H}$ is the identity mapping and $v(t,\omega;v_0)$ denotes the solution of Eq. (3.42) emanating from v_0 at $t = 0$ in the fiber ω.

Since L_λ generates an analytic semigroup e^{tL_λ} on $\mathscr{H}$, Eq. (3.43) has a unique solution for each $v_0 \in \mathscr{H}$, so does Eq. (3.42) and its corresponding solution is given by:

$$v(t,\omega;v_0) = e^{\int_0^t z_\sigma(\theta_\tau\omega)\,\mathrm{d}\tau \mathrm{Id}} e^{tL_\lambda} v_0, \quad \forall\, t \geq 0,\ v_0 \in \mathscr{H}. \tag{3.44}$$

From this observation, we can associate Eq. (3.42) with a solution operator

$$\mathfrak{T}_{\lambda,\sigma}(t_f, t_s; \omega)\colon \mathscr{H} \to \mathscr{H}$$

given by:

$$\begin{aligned}\mathfrak{T}_{\lambda,\sigma}(t_f, t_s; \omega)v_0 &:= v(t_f - t_s, \theta_{t_s}\omega; v_0) = e^{\int_0^{t_f - t_s} z_\sigma(\theta_\tau\theta_{t_s}\omega)\,\mathrm{d}\tau \mathrm{Id}} e^{(t_f - t_s)L_\lambda} v_0 \\ &= e^{\int_{t_s}^{t_f} z_\sigma(\theta_\tau\omega)\,\mathrm{d}\tau \mathrm{Id}} e^{(t_f - t_s)L_\lambda} v_0, \quad \forall\ t_s \leq t_f,\ v_0 \in \mathscr{H}.\end{aligned} \tag{3.45}$$

Similar to e^{tL_λ}, the solution operator $\mathfrak{T}_{\lambda,\sigma}(t_f, t_s; \omega)$ leaves invariant the subspaces $\mathscr{H}^{\mathfrak{c}}$ and $\mathscr{H}^{\mathfrak{s}}$; and $\mathfrak{T}_{\lambda,\sigma}(t_f, t_s; \omega)P_{\mathfrak{c}}$ can be defined for $t_f < t_s$ since $L_\lambda^{\mathfrak{c}}$ is bounded on $\mathscr{H}^{\mathfrak{c}}$.

From these simple facts about $\mathfrak{T}_{\lambda,\sigma}(t_f, t_s; \omega)$, we can easily derive the partial-dichotomy estimates for $\mathfrak{T}_{\lambda,\sigma}(t_f, t_s; \omega)$ from those associated with e^{tL_λ} given in (3.24a)–(3.24c), which are summarized in the following:

Lemma 3.2 *Let $P_{\mathfrak{c}}$ and $P_{\mathfrak{s}}$ be the projectors given in* (3.21), *$\mathfrak{T}_{\lambda,\sigma}$ be the solution operator introduced above, and η_1 and η_2 be two constants satisfying* (3.23). *Then, for each ω the following estimates hold:*

$$\|\mathfrak{T}_{\lambda,\sigma}(t_f, t_s; \omega)P_{\mathfrak{s}}\|_{L(\mathscr{H}_\alpha, \mathscr{H}_\alpha)} \leq K e^{\eta_2(t_f - t_s) + \int_{t_s}^{t_f} z_\sigma(\theta_\tau\omega)\,\mathrm{d}\tau}, \quad t_s \leq t_f, \tag{3.46a}$$

$$\|\mathfrak{T}_{\lambda,\sigma}(t_f, t_s; \omega)P_{\mathfrak{s}}\|_{L(\mathscr{H}, \mathscr{H}_\alpha)} \leq \frac{K}{(t_f - t_s)^\alpha} e^{\eta_2(t_f - t_s) + \int_{t_s}^{t_f} z_\sigma(\theta_\tau\omega)\,\mathrm{d}\tau}, \quad t_s < t_f, \tag{3.46b}$$

$$\|\mathfrak{T}_{\lambda,\sigma}(t_f, t_s; \omega)P_{\mathfrak{c}}\|_{L(\mathscr{H}, \mathscr{H}_\alpha)} \leq K e^{\eta_1(t_f - t_s) - \int_{t_f}^{t_s} z_\sigma(\theta_\tau\omega)\,\mathrm{d}\tau}, \quad t_f \leq t_s. \tag{3.46c}$$

Remark 3.4 Obviously, $e^{-\int_0^t z_\sigma(\theta_\tau\omega)\,\mathrm{d}\tau \mathrm{Id}}v = e^{-\int_0^t z_\sigma(\theta_\tau\omega)\,\mathrm{d}\tau}v$ for any $v \in \mathcal{H}$. We kept the operator Id in the above presentation, to emphasize that the linear multiplicative noise that we consider here acts "diagonally". The material of this monograph could have been adapted to address the case of more general multiplicative noise such as $Mu \circ \mathrm{d}W_t$, with W_t a space-time white noise and M a bounded Hilbert-Schmidt linear operator; see [58, 124]. We restricted our attention to the case $M = \mathrm{Id}$ and W_t one-dimensional, to present in a simple setting the main ideas of the proof of Theorem 6.1, as well as about the pullback characterization of approximating manifolds presented in Volume II [37, Sect. 4.1], and the notion of stochastic parameterizing manifolds introduced therein [37, Sect. 4.2].

Chapter 4
Existence and Attraction Properties of Global Stochastic Invariant Manifolds

In this chapter, we revisit the existence and attraction properties of families of global stochastic invariant manifolds for the SPDE (3.1) in order to set up the precise framework which we will rely on to present the main results of this first volume dealing with approximation formulas of stochastic critical manifolds (Chap. 6), and to describe their related pullback characterizations addressed in Volume II [37, Sect. 4.1].

Let us first recall the transformed RPDE associated with the SPDE (3.1):

$$\frac{\mathrm{d}u}{\mathrm{d}t} = L_\lambda u + z_\sigma(\theta_t\omega)u + G(\theta_t\omega,\, u), \tag{4.1}$$

where

$$G(\omega,\, u) = e^{-z_\sigma(\omega)} F(e^{z_\sigma(\omega)}u). \tag{4.2}$$

Following a classical approach [42, 46, 66, 152–154], we will make use—for the existence theory of random invariant manifolds—of standard weighted Banach spaces C_η^- given by:

$$\begin{aligned} C_\eta^-(\omega) := \Big\{\phi\colon (-\infty,\, 0] \to \mathscr{H}_\alpha \mid \phi \text{ is continuous, and} \\ \sup_{t\le 0} e^{-\eta t + \int_t^0 z_\sigma(\theta_\tau\omega)\,\mathrm{d}\tau} \|\phi(t)\|_\alpha < \infty\Big\}, \quad \eta \in \mathbb{R},\ \omega \in \Omega, \end{aligned} \tag{4.3}$$

endowed with the norm:

$$\|\phi\|_{C_\eta^-(\omega)} := \sup_{t\le 0} e^{-\eta t + \int_t^0 z_\sigma(\theta_\tau\omega)\,\mathrm{d}\tau} \|\phi(t)\|_\alpha. \tag{4.4}$$

Similar spaces will also be used in the study of attraction properties of stochastic invariant manifolds in Sect. 4.2.

Throughout this monograph, we suppress the ω-dependence of the space C_η^- for simplicity, since for most of the results derived in this monograph, we work

M.D. Chekroun et al., *Approximation of Stochastic Invariant Manifolds*,
SpringerBriefs in Mathematics, DOI 10.1007/978-3-319-12496-4_4

within a fixed fiber ω. Only one exception is made in the Proof of Theorem 4.1 (about existence of random invariant manifolds) where the ω-dependence is specified explicitly to show the invariance property of the global random invariant manifolds which involves two fibers *per se*; see Step 4 of the proof provided in Appendix B.

Recall that a set valued mapping $\mathfrak{C}\colon \Omega \to 2^H$ taking values in the closed (resp. compact) subsets of a separable Hilbert space $(H, |\cdot|_H)$ is called a *random closed (resp. compact) set* if for each $\xi \in H$, the map $\omega \to \mathrm{dist}(\xi, \mathfrak{C}(\omega))$ is measurable; see, e.g., [54, Definition 2.1]. Here and throughout the monograph, $\mathrm{dist}(\cdot, \cdot)$ denotes the Hausdorff semi-distance associated with the underlying space $(H, |\cdot|_H)$, i.e.,

$$\mathrm{dist}(D, E) := \sup_{a\in D} \inf_{b\in E} |a - b|_H, \ \forall\, D, E \subset H. \tag{4.5}$$

We will also make use of the following definition of a forward random invariant set of an RDS; see [54, Definition 6.9].

Definition 4.1 Let S be a continuous RDS acting on some separable Hilbert space H over some MDS $(\Omega, \mathcal{F}, \mathbb{P}, \{\theta_t\}_{t\in\mathbb{R}})$. A random closed set $\mathfrak{B}$ is said to be forward invariant for this RDS if:

$$S(t, \omega)\mathfrak{B}(\omega) \subset \mathfrak{B}(\theta_t\omega), \quad \forall\, t > 0, \ \omega \in \Omega. \tag{4.6}$$

If the inclusion in (4.6) is replaced by a set equality, the random set $\mathfrak{B}$ is said to be forward strictly invariant.

Let us now introduce the following definition of a random (resp. stochastic) invariant manifold:

Definition 4.2 Let S be a continuous RDS acting on some separable Hilbert space H. A random closed set $\mathfrak{M}$ is called a global random invariant Lipschitz (resp. C^r with $r \geq 1$) manifold of S if the following conditions are satisfied:

(i) $\mathfrak{M}$ is invariant in the sense of Definition 4.1.
(ii) There exist a closed subspace $\mathcal{G} \subset H$ and a measurable mapping $h\colon \mathcal{G} \times \Omega \to \mathcal{L}$, with $\mathcal{L}$ the topological complement of $\mathcal{G}$ in H, such that $\mathfrak{M}(\omega)$ can be represented as the graph of $h(\cdot, \omega)$ for each $\omega \in \Omega$, i.e.,

$$\mathfrak{M}(\omega) = \{\xi + h(\xi, \omega) \mid \xi \in \mathcal{G}\}, \quad \omega \in \Omega.$$

(iii) For each ω, $h(\cdot, \omega)$ is Lipschitz and $h(0, \omega) = 0$. In the case where $\mathfrak{M}$ is C^r then $h(\cdot, \omega)$ is furthermore C^r and $\mathfrak{M}$ is tangent to $\mathcal{G}$ at the origin, i.e.,

$$h(0, \omega) = 0 \qquad \text{and} \qquad D_\xi h(\xi, \omega)\big|_{\xi=0} = 0,$$

where $D_\xi h$ denotes the Fréchet derivative of h with respect to ξ.

The function h is called the global random invariant manifold function associated with $\mathfrak{M}$.

In the case where the RDS is associated with a stochastic evolution equation of type (3.1), we will refer such a manifold as a stochastic invariant Lipschitz (resp. C^r) manifold.

4.1 Existence and Smoothness of Global Stochastic Invariant Manifolds

In this section, we report on results concerning the existence and smoothness of families of global stochastic (resp. random) invariant manifolds for the SPDE (3.1) (resp. the RPDE (4.1)). These results are mainly known, but are framed here with respect to the uniform spectrum decomposition (3.11) of the linear part. In particular, the analysis is tailored to make sure the dependence on λ remains explicit.

The approach adopted here is classical but to make the expository as much self-contained as possible, a proof of the existence theorem below is provided in Appendix B. The inclusion of this proof is also motivated by the fact that some of its elements are used in the Proofs of Theorem 4.3 and of Theorem 6.1; the latter constituting the main result of this monograph, regarding the approximation formulas of stochastic critical manifolds.

More specifically, the proofs of the aforementioned theorems are rooted in the Lyapunov-Perron approach [19, 109, 114, 131], and is based on techniques used for instance in [46, 92, 147, 152–154], from which we propose a treatment adapted to the random setting that follows mainly the works of [42, 66].

Theorem 4.1 *Consider the RPDE* (4.1)*. The assumptions on L_λ and F are those of Sect.* 3.1 *where F is assumed to be globally Lipschitz with Lipschitz constant* $\mathrm{Lip}(F)$ *as given in* (3.7)*. We assume that an open interval Λ is chosen such that the uniform spectrum decomposition* (3.11) *holds over Λ, and that there exist η_1 and η_2 as in* (3.23) *for which the following uniform spectral gap condition holds:*

$$\boxed{\begin{aligned}&\exists\ \eta \in (\eta_2, \eta_1) \quad \text{s.t.}\\ &\qquad \Upsilon_1(F) := K\mathrm{Lip}(F)\big((\eta_1-\eta)^{-1} + \Gamma(1-\alpha)(\eta-\eta_2)^{\alpha-1}\big) < 1,\end{aligned}} \tag{4.7}$$

where K is the constant given in the partial-dichotomy estimates (3.24)*, and $\Gamma(s) := \int_0^\infty t^{s-1}e^{-t}\,\mathrm{d}t$ is the Gamma function. Let the spaces $\mathcal{H}^{\mathfrak{c}}$ (such that* $\dim(\mathcal{H}^{\mathfrak{c}}) = m$*) and $\mathcal{H}^{\mathfrak{s}}_\alpha$ be the corresponding subspaces associated with the uniform spectrum decomposition as defined in* (3.18) *and* (3.20)*, respectively.*

Then, for each $\lambda \in \Lambda$, there exists a global random invariant Lipschitz manifold $\mathfrak{M}_\lambda$ of the RDS, S_λ, associated with Eq. (4.1)*.*

Each of such manifolds is m-dimensional and is given by

$$\mathfrak{M}_\lambda(\omega) := \{\xi + h_\lambda(\xi, \omega) \mid \xi \in \mathscr{H}^{\mathfrak{c}}\}, \quad \omega \in \Omega, \ \lambda \in \Lambda,$$

where, for each $\lambda \in \Lambda$, $h_\lambda : \mathscr{H}^{\mathfrak{c}} \times \Omega \to \mathscr{H}_\alpha^{\mathfrak{s}}$ *is the solution of the following integral equation*[1]*:*

$$h_\lambda(\xi, \omega) = \int_{-\infty}^{0} e^{\int_s^0 z_\sigma(\theta_\tau \omega)\, d\tau \,\mathrm{Id}} e^{-sL_\lambda} P_{\mathfrak{s}} G\big(\theta_s \omega, u_\lambda(s, \omega; \xi + h_\lambda(\xi, \omega))\big)\, ds, \qquad (4.8)$$

with $u_\lambda(\cdot, \omega; \xi + h_\lambda(\xi, \omega)) \in C_\eta^-$ *being the mild solution*[2] *of Eq.* (4.1) *on* $(-\infty, 0]$ *verifying* $u_\lambda(0, \omega; \xi + h_\lambda(\xi, \omega)) = \xi + h_\lambda(\xi, \omega)$.

Furthermore, $h_\lambda(\cdot, \omega) : \mathscr{H}^{\mathfrak{c}} \to \mathscr{H}_\alpha^{\mathfrak{s}}$ *depends continuously on* λ*, and its Lipschitz constant satisfies*

$$\mathrm{Lip}(h_\lambda(\cdot, \omega)) \le \frac{K^2 \mathrm{Lip}(F)(\eta - \eta_2)^{\alpha - 1}\Gamma(1 - \alpha)}{1 - \Upsilon_1(F)}, \quad \omega \in \Omega. \qquad (4.9)$$

Proof See Appendix B. □

Since the Lipschitz constant of $h_\lambda(\cdot, \omega)$ admits a deterministic upper bound for each ω as seen in (4.9), for the sake of concision, we will denote it by $\mathrm{Lip}(h_\lambda)$ hereafter.

Remark 4.1

(1) The random invariant manifolds, $\mathfrak{M}_\lambda$, as provided by the above theorem, allow us to parametrize the high modes (in $\mathscr{H}_\alpha^{\mathfrak{s}}$) of the solutions of Eq. (4.1)—evolving on $\mathfrak{M}_\lambda$—by their low modes (in $\mathscr{H}^{\mathfrak{c}}$). However in the general case, $\mathfrak{M}_\lambda$ does not provide such a parameterization for all the solutions of Eq. (4.1). It turns out indeed that in order those manifolds to be genuine random (resp. stochastic) inertial manifolds, extra assumptions on the spectral gap condition (4.7) are (as usual) required; see Theorem 4.3 (resp. Corollary 4.3).
(2) An alternative class of useful manifolds to deal with the parameterization problem of the unresolved modes by the resolved ones, will be introduced in Volume II [37]. These manifolds, called stochastic *parameterizing manifolds* (PMs), provide an approximate parameterization of the unresolved variables by the resolved ones in a mean square sense. In contrast to the classical theory, certain PMs (in the self-adjoint case) can be determined under a *non-resonance conditions* which circumvent the constraints inherent to standard spectral gap conditions (cf. [37, Sect. 4.3] and the (NR2)-condition in [37, Sect. 7.3]); see also [37, Theorems 4.2–4.4 and Corollary 4.1].

[1] Note that Id denotes here the identity mapping of $\mathscr{H}$.

[2] In the sense of Definition A.1 provided in Appendix A.

We provide in the remark below, useful interpretations from a dynamical perspective of $\mathfrak{M}_\lambda$ such as obtained via Theorem 4.1.

Remark 4.2

(1) When $\eta_2 < \eta_1 < 0$, the Proof of Theorem 4.1 given in Appendix B shows that $\mathfrak{M}_\lambda(\omega)$ consists of all elements in $\mathcal{H}_\alpha$ such that there exists a complete trajectory of Eq. (4.1) passing through each such element at $t = 0$, which has a controlled growth rate as $t \to -\infty$. More precisely, the $\mathcal{H}_\alpha$-norm of such a solution is controlled by $e^{\eta t - \int_t^0 z_\sigma(\theta_\tau\omega)\,\mathrm{d}\tau}$ as $t \to -\infty$, where $\eta \in (\eta_2, \eta_1)$ is chosen such that the condition (4.7) holds, and $z_\sigma(\theta_t\omega)$ is the OU process defined in (3.29). Hyperbolic manifolds can be encountered in this case; see Chap. 7. See also [37, Chaps. 6, 7] where hyperbolic manifolds are dealt with in details in the context of a stochastic Burgers-type equation.
(2) When $\eta_2 < 0 < \eta_1$ in (3.24), we are in the case of a classical dichotomy. The dynamical interpretation of $\mathfrak{M}_\lambda$ for this case is more standard. Indeed, if η in (4.7) is positive, then $\mathfrak{M}_\lambda$ can be interpreted as the random closed set consisting of all elements in $\mathcal{H}_\alpha$ such that the corresponding mild solutions decay to 0 as $t \to -\infty$ since in this case, $e^{\eta t - \int_t^0 z_\sigma(\theta_\tau\omega)\,\mathrm{d}\tau} \to 0$ as $t \to -\infty$. The manifold $\mathfrak{M}_\lambda$ is just the random unstable manifold of the trivial steady state.
(3) By adapting the Proof of Theorem 4.1 accordingly, it can be proved that for each $\lambda \in \Lambda$ the RDS associated with Eq. (4.1) admits also a random invariant manifold given by

$$\mathfrak{M}_\lambda^{\mathfrak{s}}(\omega) = \{\zeta + h_\lambda(\zeta, \omega) \mid \zeta \in \mathcal{H}_\alpha^{\mathfrak{s}}\}, \quad \omega \in \Omega, \tag{4.10}$$

where the invariant manifold function $h_\lambda(\cdot, \omega)$ is here defined over $\mathcal{H}_\alpha^{\mathfrak{s}}$ with, this time, range in $\mathcal{H}^{\mathfrak{c}}$.
If in addition the classical dichotomy is satisfied, and η in (4.7) is negative, then $\mathfrak{M}_\lambda^{\mathfrak{s}}$ given in (4.10) corresponds to a random stable manifold which can be interpreted as the random closed set consisting of all initial data through which the mild solutions converges to 0 as $t \to +\infty$. We will not make use of such manifolds in this monograph.

Concerning the smoothness of the random invariant manifold $\mathfrak{M}_\lambda$ given by Theorem 4.1, we have the following

Theorem 4.2 *Consider the RPDE* (4.1). *Assume that the assumptions of Theorem* 4.1 *hold with Λ specified therein. The nonlinearity $F : \mathcal{H}_\alpha \to \mathcal{H}$ is furthermore assumed to be C^p-smooth for some integer $p \geq 1$ and to satisfy $DF(0) = 0$. Assume also that there exist $\eta \in \mathbb{R}$ and some integer $r \in \{1, \dots, p\}$, such that*

$$\eta_2 < j\eta < \eta_1, \quad \forall\, j \in \{1, \dots, r\}, \tag{4.11}$$

and the following uniform spectral gap conditions hold

$$\boxed{\Upsilon_j(F) := K\mathrm{Lip}(F)\big((\eta_1-\eta)^{-1}+\Gamma(1-\alpha)(j\eta-\eta_2)^{\alpha-1}\big) < 1, \quad \forall\, j \in \{1,\ldots,r\}.} \tag{4.12}$$

Then, for each $\lambda \in \Lambda$, the random invariant manifold $\mathfrak{M}_\lambda$ obtained in Theorem 4.1 *is C^r-smooth. In particular, the random invariant manifold function, $(\xi,\omega) \mapsto h_\lambda(\xi,\omega)$, guaranteed therein, is C^r-smooth with respect to ξ for each ω, and is tangent to the subspace $\mathcal{H}^{\mathrm{c}}$ at the origin, i.e., $h_\lambda(0,\omega) = 0$, and $D_\xi h_\lambda(\xi,\omega)|_{\xi=0} = 0$.*

Proof The proof follows essentially the same lines of the proof of [66, Theorem 4.1] and is omitted here. □

By the cohomology relation (3.41) between the RDSs associated respectively with Eqs. (3.1) and (4.1), we obtain the following corollaries regarding the existence and smoothness of families of stochastic invariant manifolds $\widehat{\mathfrak{M}}_\lambda$ for Eq. (3.1).

Corollary 4.1 *Consider the SPDE* (3.1). *Assume that the assumptions of Theorem* 4.1 *hold with Λ specified therein. Let also $h_\lambda : \mathcal{H}^{\mathrm{c}} \times \Omega \to \mathcal{H}^{\mathrm{s}}_\alpha$ be the random invariant manifold function associated with Eq.* (4.1) *provided by Theorem* 4.1. *Then, for each $\lambda \in \Lambda$, the RDS, $\widehat{S}_\lambda$, as defined in* (3.41) *possesses an m-dimensional global stochastic invariant Lipschitz manifold $\widehat{\mathfrak{M}}_\lambda$ given by:*

$$\widehat{\mathfrak{M}}_\lambda(\omega) := \{\xi + e^{z_\sigma(\omega)} h_\lambda(e^{-z_\sigma(\omega)}\xi, \omega) \mid \xi \in \mathcal{H}^{\mathrm{c}}\}, \quad \omega \in \Omega. \tag{4.13}$$

Corollary 4.2 *Consider the SPDE* (3.1). *Assume that the assumptions of Theorem* 4.2 *hold with r specified therein. Then for each $\lambda \in \Lambda$, $\widehat{\mathfrak{M}}_\lambda$ defined in* (4.13) *is a global stochastic invariant C^r-manifold of $\widehat{S}_\lambda$.*

4.2 Asymptotic Completeness of Stochastic Invariant Manifolds

We establish in this section the attraction properties of random invariant manifolds as stated in Theorem 4.3 below. As explained in Chap. 2, the corresponding results may be viewed as complementary to previous results obtained on the topic [12, 42, 51, 57, 142, 144, 155]. Furthermore, some new insights concerning the asymptotic completeness problem are provided. Our approach is inspired by [46, Theorem 5.1] that we adapt to our framework; see also [42].

More precisely, for a given solution u_λ to Eq. (4.1) we look for a solution $\overline{u}_\lambda$ living on the random invariant manifold $\mathfrak{M}_\lambda$ such that $\|\overline{u}_\lambda(t,\omega) - u_\lambda(t,\omega)\|_\alpha$ decays exponentially as $t \to \infty$ for almost all ω. The strategy adopted here consists of reformulating this problem as a fixed point problem under the constraint that the sought solution $\overline{u}_\lambda$ emanates from an initial datum $\overline{u}_0$ which belongs to $\mathfrak{M}_\lambda$ and is well-prepared with respect to the given initial datum u_0.

This problem is then recast as an unconstrained fixed point problem associated with a random integral operator that is solved by means of the uniform contraction mapping principle [44, Theorems 2.1 and 2.2]. The price to pay is an additional condition (4.16) to the existence theory which involves the spectral gap and the Lipschitz constant associated with $\mathfrak{M}_\lambda$.

The results so obtained show that the manifolds obtained in Sect. 4.1 under this additional condition are, for $\eta < 0$, almost surely forward and pullback asymptotically complete at a *uniform rate* with respect to the parameter λ. By doing so, we establish naturally the existence of stochastic inertial manifolds, which are almost surely exponentially attracting at a uniform rate in both a forward and a pullback sense. Before presenting the results, let us introduce the following definition of forward (resp. pullback) asymptotic completeness of a stochastic invariant manifold, which is motivated by its deterministic analogue; see, e.g., [52, 137].

Definition 4.3 Let S be a continuous RDS acting on some separable Hilbert space H over some MDS, $(\Omega, \mathscr{F}, \mathbb{P}, \{\theta_t\}_{t\in\mathbb{R}})$, and $\mathfrak{M}$ be a random invariant manifold of S. The manifold $\mathfrak{M}$ is said to be forward asymptotically complete if there exist $\kappa > 0$ and a positive random variable C_κ such that for any H-valued tempered random initial datum[3] u_0 there exists a tempered random variable v_0 on the manifold $\mathfrak{M}$ satisfying

$$\begin{aligned} &\|S(t,\omega)u_0(\omega) - S(t,\omega)v_0(\omega)\| \\ &\quad \le C_\kappa(\omega)\|u_0(\omega) - v_0(\omega)\|e^{-\kappa t}, \quad t \ge 0,\ \omega \in \Omega. \end{aligned} \tag{4.14}$$

It is said to be pullback asymptotically complete if under the same conditions the following attraction property holds:

$$\begin{aligned} &\|S(t,\theta_{-t}\omega)u_0(\theta_{-t}\omega) - S(t,\theta_{-t}\omega)v_0(\theta_{-t}\omega)\| \\ &\quad \le C_\kappa(\omega)\|u_0(\theta_{-t}\omega) - v_0(\theta_{-t}\omega)\|e^{-\kappa t}, \quad t \ge 0,\ \omega \in \Omega. \end{aligned} \tag{4.15}$$

The rate κ is called an attraction rate. The supremum of such attraction rates is called the critical attraction rate.[4]

If furthermore $\mathfrak{M}(\omega)$ is finite-dimensional of fixed dimension for all ω, then $\mathfrak{M}$ is called a pullback (resp. forward) random inertial manifold in case where (4.15) (resp. (4.14)) holds.

Theorem 4.3 *Consider the RPDE* (4.1). *Assume that the assumptions of Theorem* 4.1 *hold with Λ and $\Upsilon_1(F)$ specified therein. Assume also that η in condition* (4.7) *is negative. If the invariant manifold function h_λ guaranteed by Theorem* 4.1 *satisfies furthermore*

$$\frac{K\mathrm{Lip}(h_\lambda)}{1-\Upsilon_1(F)} < 1, \tag{4.16}$$

then for each $\lambda \in \Lambda$ the random invariant manifold $\mathfrak{M}_\lambda$ obtained by Theorem 4.1 *is both a pullback and a forward random inertial manifold with critical attraction rate $|\eta|$.*

---

[3] See, e.g., [47, Definition 1.3.3] for the definition of a tempered random variable.

[4] Note that such supremum is not an attraction rate in general.

Once more, by using the cohomology relation between the two RDSs associated respectively with Eqs. (3.1) and (4.1), we obtain the following results on the forward and pullback asymptotic completeness of the stochastic invariant manifolds $\widehat{\mathfrak{M}}_\lambda$ for Eq. (3.1) as provided by Corollary 4.1.

Corollary 4.3 *Consider the SPDE* (3.1). *Assume that the assumptions of Theorem* 4.3 *hold with Λ and η specified therein. Then for each $\lambda \in \Lambda$ the stochastic invariant manifold $\widehat{\mathfrak{M}}_\lambda$ guaranteed by Corollary* 4.1 *is both a pullback and a forward stochastic inertial manifold with critical attraction rate $|\eta|$.*

Remark 4.3 Results similar to Theorem 4.3 in the global setting, misled some authors to conclude about *pathwise* asymptotic completeness results for *local* stochastic invariant manifolds such as ensured by e.g., Corollary 5.1 and which are defined as a graph of random *Lipschitz* function over a deterministic bounded neighborhood $\mathscr{V} \subset \mathscr{H}^{\mathfrak{c}}$ of the basic state. Indeed if one denotes by γ the deterministic upper bound in (4.9) (see Theorem 4.1), the latter inequality implies that the corresponding manifold function satisfies

$$\|h_\lambda(\xi, \omega)\|_\alpha \leq \gamma \operatorname{diam}(\mathscr{V}), \omega \in \Omega, \xi \in \mathscr{V},$$

which would imply that for any solution $u \neq 0$ of (3.1), $\operatorname{dist}(u(t, \omega), \mathfrak{M}(\omega))$ remains uniformly bounded as t flows (with respect to a.e. ω) if one assume the pathwise attraction property (4.14) to hold. Such a boundedness property is clearly in contradiction with the large deviations principles [48, 115] demonstrating thus that (4.14) cannot hold almost surely for local stochastic invariant manifolds such as discussed in this remark. In particular [41, Lemma 9] should be reconsidered in a probabilistic sense.

We first present the main ideas of the proof.

Ideas of the Proof of Theorem 4.3. As mentioned above, we describe here what fixed point problem under constraint is naturally related to the problem of asymptotic completeness and how this problem can be solved by recasting it as an unconstrained fixed point problem.

In that respect, we first point out in Step 1 a natural integral equation to be satisfied by the difference, $v_\lambda := \overline{u}_\lambda - u_\lambda$, between any solution $\overline{u}_\lambda$ of Eq. (4.1) and a solution u_λ emanating from a given $\mathscr{H}_\alpha$-valued tempered random initial datum u_0:

$$\begin{aligned} v_\lambda(t, \omega) &= \mathfrak{T}_{\lambda,\sigma}(t, t_0; \omega) v_\lambda(t_0, \omega) \\ &\quad + \int_{t_0}^{t} \mathfrak{T}_{\lambda,\sigma}(t, s; \omega) \delta G(\theta_s \omega, v_\lambda(s, \omega))\, ds, \quad 0 \leq t_0 < t, \ \omega \in \Omega, \end{aligned}$$

where $\delta G(\theta_s\omega, v_\lambda(s, \omega)) = G(\theta_s\omega, u_\lambda(s, \omega) + v_\lambda(s, \omega)) - G(\theta_s\omega, u_\lambda(s, \omega))$, and $\mathfrak{T}_{\lambda,\sigma}$ is the linearized stochastic flow introduced in Sect. 3.4.

The above integral equation is then rewritten into the following fixed point problem:

$$v_\lambda(t,\omega) = \mathscr{L}_q^{\omega,\lambda}[v_\lambda](t), \quad t \geq 0, \tag{4.17}$$

where

$$\begin{aligned}\mathscr{L}_q^{\omega,\lambda}[v_\lambda](t) := \mathfrak{T}_{\lambda,\sigma}(t,0;\omega)q + \int_0^t \mathfrak{T}_{\lambda,\sigma}(t,s;\omega)P_{\mathfrak{s}}\delta G(\theta_s\omega, v_\lambda(s,\omega))\,\mathrm{d}s \\ - \int_t^{+\infty} \mathfrak{T}_{\lambda,\sigma}(t,s;\omega)P_{\mathfrak{c}}\delta G(\theta_s\omega, v_\lambda(s,\omega))\,\mathrm{d}s, \quad t \geq 0,\end{aligned} \tag{4.18}$$

with $q = P_{\mathfrak{s}}v_\lambda(0,\omega)$. The advantage of doing so is to dispose of an operator $\mathscr{L}_q^{\omega,\lambda}$ which is written under a form suitable for establishing its contraction property based on the partial-dichotomy estimates (3.46); see Steps 2 and 3.

Now, given $v_\lambda[q]$ which solves (4.17), since $\overline{u}_\lambda = v_\lambda[q] + u_\lambda$, in order to show that $\overline{u}_\lambda$ lives on $\mathfrak{M}_\lambda$, it is sufficient to show that there exists $q \in \mathscr{H}_\alpha^{\mathfrak{s}}$ such that $v_\lambda[q](0,\omega) + u_0(\omega) \in \mathfrak{M}_\lambda(\omega)$ for each $\omega \in \Omega$, due to the invariance property of $\mathfrak{M}_\lambda$. In that respect, we introduce a weighted Banach space C_η^+,[5]

$$C_\eta^+ := \{\phi\colon [0,\infty) \to \mathscr{H}_\alpha \mid \phi \text{ is continuous and } \sup_{t\geq 0} e^{-\eta t - \int_0^t z_\sigma(\theta_\tau\omega)\,\mathrm{d}\tau}\|\phi(t)\|_\alpha < \infty\}, \tag{4.19}$$

which is endowed with the following norm:

$$|\phi|_{C_\eta^+} := \sup_{t\geq 0} e^{-\eta t - \int_0^t z_\sigma(\theta_\tau\omega)\,\mathrm{d}\tau}\|\phi(t)\|_\alpha. \tag{4.20}$$

If furthermore we show that $v_\lambda[q](\cdot,\omega)$ can be obtained for each ω as a fixed point of $\mathscr{L}_q^{\omega,\lambda}$ in the space C_η^+ for some $\eta < 0$ (independent of ω) chosen according to (4.7), then we get naturally that $v_\lambda[q](t,\omega) = \overline{u}_\lambda(t,\omega) - u_\lambda(t,\omega)$ approaches 0 exponentially as $t \to \infty$, so that the asymptotic completeness problem is solved.

The problem of searching for the desired solution $\overline{u}_\lambda$ which solves the asymptotic completeness problem is then recast into the problem of searching for a fixed point $v_\lambda[q](\cdot,\omega)$ of the integral operator $\mathscr{L}_q^{\omega,\lambda}$ given by (4.18) in the weighted Banach space C_η^+ under the constraint that $\overline{u}_0(\omega) := v_\lambda[q](0,\omega) + u_0(\omega) \in \mathfrak{M}_\lambda(\omega)$ for each $\omega \in \Omega$, namely, the following problem

$$\begin{cases} \mathscr{L}_q^{\omega,\lambda}[v] = v, \quad v \in C_\eta^+, \quad \eta < 0, \\ v(0,\omega) + u_0(\omega) \in \mathfrak{M}_\lambda(\omega), \quad \omega \in \Omega. \end{cases} \tag{4.21}$$

[5] We suppress the ω-dependence of C_η^+ for the same reasons that were invoked for C_η^- in (4.3).

The problem (4.21) is solved in two stages.

In the first stage, we deal with the fixed point problem $\mathscr{L}_q^{\omega,\lambda}[v] = v$ (in C_η^+) disregarding first the constraint, which is done in Steps 2 and 3. It is shown in Step 2 that $\mathscr{L}_q^{\omega,\lambda}$ leaves C_η^+ stable, and in Step 3 that $\mathscr{L}_q^{\omega,\lambda}$ has a unique fixed point $v_\lambda[q](\cdot,\omega) \in C_\eta^+$ parameterized by $q = P_{\mathfrak{s}} v_\lambda[q](0,\omega) \in \mathscr{H}_\alpha^{\mathfrak{s}}$ using the uniform contraction mapping principle [44, Theorems 2.1 and 2.2]. Also derived in Step 3 is a control of the C_η^+-norm of $v_\lambda[q](\cdot,\omega)$ in terms of $\|q\|_\alpha$ which is uniform with respect to ω; see (4.36). This will turn out to be important to ensure the temperedness of the initial datum $\overline{u}_0$ examined in Step 5.

In the second stage, with the preparation carried out in the first stage we solve the constrained fixed point problem (4.21) by transforming it into a fixed point problem without constraint. This is done in Step 4. First note that given a $q \in \mathscr{H}_\alpha^{\mathfrak{s}}$ and the corresponding fixed point $v_\lambda[q](\cdot,\omega)$ of $\mathscr{L}_q^{\omega,\lambda}$, an initial datum $\overline{u}_0(\omega)$ given by $v_\lambda[q](0,\omega) + u_0(\omega)$ belongs to $\mathfrak{M}_\lambda(\omega)$ if $\overline{u}_0(\omega) = p + h_\lambda(p,\omega)$ for some $p \in \mathscr{H}^{\mathfrak{c}}$, where h_λ is the random invariant manifold function of Eq. (4.1) as provided by Theorem 4.1. Since $q = P_{\mathfrak{s}} v_\lambda[q](0,\omega)$, it is thus natural to seek for q of the following form:

$$q = P_{\mathfrak{s}}(\overline{u}_0(\omega) - u_0(\omega)) = h_\lambda(p,\omega) - P_{\mathfrak{s}} u_0(\omega).$$

The constraint $\overline{u}_0(\omega) \in \mathfrak{M}_\lambda(\omega)$ is then written as a functional equation in terms of p:

$$\begin{aligned} p = P_{\mathfrak{c}}\overline{u}_0(\omega) &= P_{\mathfrak{c}} v_\lambda[q](0,\omega) + P_{\mathfrak{c}} u_0(\omega) \\ &= P_{\mathfrak{c}} v_\lambda[h_\lambda(p,\omega) - P_{\mathfrak{s}} u_0(\omega)](0,\omega) + P_{\mathfrak{c}} u_0(\omega). \end{aligned}$$

As a consequence, the problem (4.21) can be rewritten as

$$\begin{cases} \mathscr{L}_{q(p)}^{\omega,\lambda}[v] = v, \quad v \in C_\eta^+, \\ p = P_{\mathfrak{c}} v(0,\omega) + P_{\mathfrak{c}} u_0(\omega), \end{cases} \tag{4.22}$$

where $q(p) = h_\lambda(p,\omega) - P_{\mathfrak{s}} u_0(\omega)$. Note also that for the fixed point $v_\lambda[q(p)](\cdot,\omega)$ of $\mathscr{L}_{q(p)}^{\omega,\lambda}$, $P_{\mathfrak{c}} v_\lambda[q(p)](0,\omega)$ admits a natural integral representation given by:

$$\begin{aligned} &P_{\mathfrak{c}} v_\lambda[q(p)](0,\omega) \\ &= -\int_0^{+\infty} \mathfrak{T}_{\lambda,\sigma}(0,s;\omega) P_{\mathfrak{c}} \delta G(\theta_s\omega, v_\lambda[q(p)](s,\omega))\, \mathrm{d}s, \quad \forall\ p \in \mathscr{H}^{\mathfrak{c}}, \end{aligned} \tag{4.23}$$

which is obtained by applying the projector $P_{\mathfrak{c}}$ to $\mathscr{L}_{q(p)}^{\omega,\lambda}$ given in (4.18) and setting t to 0. In virtue of the integral representation (4.23), the problem (4.22) can be further transformed into the following fixed point problem in terms of p but without any constraint:

$$
\begin{aligned}
p &= -\int_0^{+\infty} \mathfrak{T}_{\lambda,\sigma}(0,s;\omega)P_{\mathfrak{c}}\delta G(\theta_s\omega, v_\lambda[q(p)](s,\omega))\,\mathrm{d}s + P_{\mathfrak{c}}u_0(\omega) \\
&=: \mathscr{J}_{\omega,\lambda}(p),
\end{aligned}
\tag{4.24}
$$

where $v_\lambda[q(p)](\cdot,\omega)$ is the fixed point of $\mathscr{L}^{\omega,\lambda}_{q(p)}$ with $q(p) = h_\lambda(p,\omega) - P_{\mathfrak{s}}u_0(\omega)$.

The unknown p is then found using the contraction mapping principle applied to the operator $\mathscr{J}_{\omega,\lambda}$ defined on $\mathscr{H}^{\mathfrak{c}}$.

In Step 5, it is shown that the random initial datum $\overline{u}_0$ obtained from the previous step is tempered by using an estimate on $v_\lambda[q](0,\omega)$ derived in Step 3 and the assumption that u_0 is tempered.

Proof of Theorem 4.3 We proceed in five steps following the ideas outlined above to prove both the forward and pullback asymptotic completeness properties of $\mathfrak{M}_\lambda$.

Step 1. Fixed point problem satisfied by v_λ. Let us first point out a natural integral equation to be satisfied by the difference, $v_\lambda := \overline{u}_\lambda - u_\lambda$, between any solution $\overline{u}_\lambda$ of Eq. (4.1) and a given solution u_λ emanating from a given $\mathscr{H}_\alpha$-valued tempered random initial datum u_0:

$$
\begin{aligned}
v_\lambda(t,\omega) &= \mathfrak{T}_{\lambda,\sigma}(t,t_0;\omega)v_\lambda(t_0,\omega) \\
&\quad + \int_{t_0}^{t} \mathfrak{T}_{\lambda,\sigma}(t,s;\omega)\delta G(\theta_s\omega, v_\lambda(s,\omega))\,\mathrm{d}s, \quad 0 \le t_0 < t,
\end{aligned}
\tag{4.25}
$$

where

$$
\delta G(\theta_s\omega, v_\lambda(s,\omega)) := G(\theta_s\omega, u_\lambda(s,\omega) + v_\lambda(s,\omega)) - G(\theta_s\omega, u_\lambda(s,\omega)), \tag{4.26}
$$

and $\mathfrak{T}_{\lambda,\sigma}$ is the solution operator defined in Sect. 3.4.

For any given $q \in \mathscr{H}^{\mathfrak{s}}_\alpha$ and $\omega \in \Omega$, let us define then a mapping $\mathscr{L}^{\omega,\lambda}_q$ acting on C^+_η as follows:

$$
\begin{aligned}
&\mathscr{L}^{\omega,\lambda}_q[v](t) \\
&:= \mathfrak{T}_{\lambda,\sigma}(t,0;\omega)q + \int_0^t \mathfrak{T}_{\lambda,\sigma}(t,s;\omega)P_{\mathfrak{s}}\delta G(\theta_s\omega, v(s,\omega))\,\mathrm{d}s \\
&\quad - \int_t^{+\infty} \mathfrak{T}_{\lambda,\sigma}(t,s;\omega)P_{\mathfrak{c}}\delta G(\theta_s\omega, v(s,\omega))\,\mathrm{d}s, \quad t \ge 0,\ v(\cdot,\omega) \in C^+_\eta,
\end{aligned}
\tag{4.27}
$$

where δG is defined in (4.26), and C^+_η is the weighted Banach space defined in (4.19).

Similar to Step 1 of the Proof of Theorem 4.1 given in Appendix B, one can show that $v_\lambda(\cdot, \omega)$ is a solution to (4.25) verifying $v_\lambda(\cdot, \omega) \in C_\eta^+$ if and only if $v_\lambda(\cdot, \omega)$ is a fixed point of the operator $\mathscr{L}_q^{\omega,\lambda}$ in C_η^+, with $q = P_{\mathfrak{s}} v_\lambda(0, \omega)$.

The rest of the proof is devoted to solving the following fixed point problem:

$$\begin{cases} \mathscr{L}_q^{\omega,\lambda}[v] = v, \quad v \in C_\eta^+, \quad \eta < 0, \\ v(0, \omega) + u_0(\omega) \in \mathfrak{M}_\lambda(\omega), \quad \omega \in \Omega, \end{cases} \tag{4.28}$$

which as explained above and in Step 5 below is sufficient to solve the asymptotic completeness problem.

Step 2. $\mathscr{L}_q^{\omega,\lambda}$ **leaves** C_η^+ **stable**. For each $q \in \mathscr{H}_\alpha^{\mathfrak{s}}$ and $\omega \in \Omega$, we show in this step that

$$\mathscr{L}_q^{\omega,\lambda} C_\eta^+ \subset C_\eta^+. \tag{4.29}$$

Since $\eta > \eta_2$ according to condition (4.7) and $q \in \mathscr{H}_\alpha^{\mathfrak{s}}$, it follows from the partial-dichotomy estimate (3.46a) that

$$\sup_{t\geq 0} e^{-\eta t - \int_0^t z_\sigma(\theta_s\omega)\, \mathrm{d}s} \|\mathfrak{T}_{\lambda,\sigma}(t, 0; \omega) q\|_\alpha \leq \sup_{t\geq 0} K e^{(\eta_2 - \eta)t} \|q\|_\alpha = K\|q\|_\alpha. \tag{4.30}$$

Recalling that G and F have the same Lipschitz constant $\mathrm{Lip}(F)$, we then derive from (4.26) that $\|\delta G(\theta_s\omega, v(s, \omega))\| \leq \mathrm{Lip}(F)\|v(s, \omega)\|_\alpha$ for any $s \geq 0$ and $v(\cdot, \omega) \in C_\eta^+$. This together with (3.46b) implies that

$$\begin{aligned} &\left\| \int_0^t \mathfrak{T}_{\lambda,\sigma}(t, s; \omega) P_{\mathfrak{s}} \delta G(\theta_s\omega, v(s, \omega))\, \mathrm{d}s \right\|_\alpha \\ &\quad \leq K \int_0^t \frac{e^{\eta_2(t-s) + \int_s^t z_\sigma(\theta_\tau\omega)\, \mathrm{d}\tau}}{(t-s)^\alpha} \|\delta G(\theta_s\omega, v(s, \omega))\|\, \mathrm{d}s \\ &\quad \leq K\mathrm{Lip}(F) \int_0^t \frac{e^{\eta_2(t-s) + \int_s^t z_\sigma(\theta_\tau\omega)\, \mathrm{d}\tau}}{(t-s)^\alpha} \|v(s, \omega)\|_\alpha\, \mathrm{d}s, \quad \forall t \geq 0,\ v(\cdot, \omega) \in C_\eta^+. \end{aligned}$$

It follows from the above inequality and the definition of the C_η^+-norm given by (4.20) that

$$
\begin{aligned}
&e^{-\eta t-\int_0^t z_\sigma(\theta_s\omega)\,\mathrm{d}s}\left\|\int_0^t \mathfrak{T}_{\lambda,\sigma}(t,s;\omega)P_{\mathfrak{s}}\delta G(\theta_s\omega, v(s,\omega))\,\mathrm{d}s\right\|_\alpha \\
&\le K\mathrm{Lip}(F)e^{-\eta t-\int_0^t z_\sigma(\theta_s\omega)\,\mathrm{d}s}\int_0^t \frac{e^{\eta_2(t-s)+\int_s^t z_\sigma(\theta_\tau\omega)\,\mathrm{d}\tau}}{(t-s)^\alpha}\|v(s,\omega)\|_\alpha\,\mathrm{d}s \\
&\le K\mathrm{Lip}(F)e^{-\eta t-\int_0^t z_\sigma(\theta_s\omega)\,\mathrm{d}s}\int_0^t \frac{e^{\eta_2(t-s)+\eta s+\int_0^t z_\sigma(\theta_\tau\omega)\,\mathrm{d}\tau}}{(t-s)^\alpha}\,\mathrm{d}s\,\|v(\cdot,\omega)\|_{C_\eta^+} \\
&= K\mathrm{Lip}(F)\int_0^t \frac{e^{(\eta_2-\eta)(t-s)}}{(t-s)^\alpha}\,\mathrm{d}s\,\|v(\cdot,\omega)\|_{C_\eta^+} \\
&\le \frac{K\mathrm{Lip}(F)\Gamma(1-\alpha)}{(\eta-\eta_2)^{1-\alpha}}\|v(\cdot,\omega)\|_{C_\eta^+}, \quad \forall\, t\ge 0,\ v(\cdot,\omega)\in C_\eta^+.
\end{aligned}
\tag{4.31}
$$

Similarly, we have for all $t \ge 0$ and $v(\cdot,\omega) \in C_\eta^+$ that

$$
\begin{aligned}
e^{-\eta t-\int_0^t z_\sigma(\theta_s\omega)\,\mathrm{d}s}&\left\|\int_t^{+\infty} \mathfrak{T}_{\lambda,\sigma}(t,s;\omega)P_{\mathfrak{c}}\delta G(\theta_s\omega, v(s,\omega))\,\mathrm{d}s\right\|_\alpha \\
&\le \frac{K\mathrm{Lip}(F)}{\eta_1-\eta}\|v(\cdot,\omega)\|_{C_\eta^+}.
\end{aligned}
\tag{4.32}
$$

Now, (4.29) follows from (4.30)–(4.32).

Step 3. Unique fixed point of $\mathscr{L}_q^{\omega,\lambda}$. In this step, we show that $\mathscr{L}_q^{\omega,\lambda}$ has a unique fixed point $v_\lambda[q](\cdot,\omega) \in C_\eta^+$ for each $q \in \mathscr{H}_\alpha^{\mathfrak{s}}$ and $\omega \in \Omega$; and that $v_\lambda[q]\colon [0,\infty)\times\Omega \to \mathscr{H}_\alpha$ is $(\mathscr{B}([0,\infty))\otimes\mathscr{F};\mathscr{B}(\mathscr{H}_\alpha))$-measurable. The Lipschitz continuity of $v_\lambda[q](t,\omega)$ with respect to q is also examined.

For any $v_1(\cdot,\omega), v_2(\cdot,\omega) \in C_\eta^+$, following the same type of estimates as given in (4.31) and (4.32), we obtain

$$
\begin{aligned}
&\|\mathscr{L}_q^{\omega,\lambda}[v_1]-\mathscr{L}_q^{\omega,\lambda}[v_2]\|_{C_\eta^+} \\
&\le \sup_{t\ge 0} e^{-\eta t-\int_0^t z_\sigma(\theta_s\omega)\,\mathrm{d}s}\left\|\int_0^t \mathfrak{T}_{\lambda,\sigma}(t,s;\omega)P_{\mathfrak{s}}\big[\delta G(\theta_s\omega,v_1)-\delta G(\theta_s\omega,v_2)\big]\,\mathrm{d}s\right\|_\alpha \\
&\quad + \sup_{t\ge 0} e^{-\eta t-\int_0^t z_\sigma(\theta_s\omega)\,\mathrm{d}s}\left\|\int_t^{+\infty} \mathfrak{T}_{\lambda,\sigma}(t,s;\omega)P_{\mathfrak{c}}\big[\delta G(\theta_s\omega,v_1)-\delta G(\theta_s\omega,v_2)\big]\,\mathrm{d}s\right\|_\alpha \\
&\le K\mathrm{Lip}(F)\big((\eta_1-\eta)^{-1}+\Gamma(1-\alpha)(\eta-\eta_2)^{\alpha-1}\big)\|v_1(\cdot,\omega)-v_2(\cdot,\omega)\|_{C_\eta^+} \\
&= \Upsilon_1(F)\|v_1(\cdot,\omega)-v_2(\cdot,\omega)\|_{C_\eta^+},
\end{aligned}
\tag{4.33}
$$

where $\Upsilon_1(F) \in (0, 1)$ is given in (4.7). This implies that $\mathscr{L}_q^{\omega,\lambda}$ is a uniform contraction in the space C_η^+ with respect to q. By the uniform contraction mapping principle, see, e.g., [44, Theorems 2.1 and 2.2], the operator $\mathscr{L}_q^{\omega,\lambda}$ has a unique fixed point $v_\lambda[q](\cdot, \omega) \in C_\eta^+$ for any $q \in \mathscr{H}_\alpha^{\mathfrak{s}}$.

The measurability of $v_\lambda[q]\colon [0, \infty) \times \Omega \to \mathscr{H}_\alpha$ can be obtained in the same fashion as in Step 2 of the proof in Appendix B, by relying on the Picard scheme naturally associated with the fixed point problem $\mathscr{L}_q^{\omega,\lambda}[v] = v$.

We turn now to the Lipschitz continuity of $v_\lambda[q]$ with respect to q. According to the definition of $\mathscr{L}_q^{\omega,\lambda}$ given in (4.27), we get

$$
\begin{aligned}
\|\mathscr{L}_{q_1}^{\omega,\lambda}[v] - \mathscr{L}_{q_2}^{\omega,\lambda}[v]\|_{C_\eta^+} &= \|\mathfrak{T}_{\lambda,\sigma}(\cdot, 0; \omega)q_1 - \mathfrak{T}_{\lambda,\sigma}(\cdot, 0; \omega)q_2\|_{C_\eta^+} \\
&= \sup_{t \geq 0} e^{-\eta t - \int_0^t z_\sigma(\theta_\tau \omega)\,\mathrm{d}\tau} \|\mathfrak{T}_{\lambda,\sigma}(t, 0; \omega)(q_1 - q_2)\|_\alpha \\
&\leq (\text{ by (4.30)}) \\
&\leq K\|q_1 - q_2\|_\alpha, \quad \forall\, q_1, q_2 \in \mathscr{H}_\alpha^{\mathfrak{s}},\ v(\cdot, \omega) \in C_\eta^+.
\end{aligned} \tag{4.34}
$$

Hence,

$$
\begin{aligned}
\|v_\lambda[q_1](\cdot, \omega) - v_\lambda[q_2](\cdot, \omega)\|_{C_\eta^+} &= \|\mathscr{L}_{q_1}^{\omega,\lambda}[v_\lambda[q_1]] - \mathscr{L}_{q_2}^{\omega,\lambda}[v_\lambda[q_2]]\|_{C_\eta^+} \\
&\leq \|\mathscr{L}_{q_1}^{\omega,\lambda}[v_\lambda[q_1]] - \mathscr{L}_{q_1}^{\omega,\lambda}[v_\lambda[q_2]]\|_{C_\eta^+} \\
&\quad + \|\mathscr{L}_{q_1}^{\omega,\lambda}[v_\lambda[q_2]] - \mathscr{L}_{q_2}^{\omega,\lambda}[v_\lambda[q_2]]\|_{C_\eta^+} \\
&\leq (\text{ by (4.33) and (4.34)}) \\
&\leq \Upsilon_1(F)\|v_\lambda[q_1](\cdot, \omega) - v_\lambda[q_2](\cdot, \omega)\|_{C_\eta^+} \\
&\quad + K\|q_1 - q_2\|_\alpha.
\end{aligned}
$$

We then obtain

$$
\|v_\lambda[q_1](\cdot, \omega) - v_\lambda[q_2](\cdot, \omega)\|_{C_\eta^+} \leq \frac{K}{1 - \Upsilon_1(F)}\|q_1 - q_2\|_\alpha, \quad \forall\, q_1, q_2 \in \mathscr{H}_\alpha^{\mathfrak{s}}. \tag{4.35}
$$

Note that when $q = 0$, $v \equiv 0$ satisfies (4.27). By the uniqueness of the fixed point, we have $v_\lambda[0](t, \omega) \equiv 0$. This together with (4.35) implies that

$$
\|v_\lambda[q](\cdot, \omega)\|_{C_\eta^+} \leq \frac{K}{1 - \Upsilon_1(F)}\|q\|_\alpha, \quad \forall\, q \in \mathscr{H}_\alpha^{\mathfrak{s}}. \tag{4.36}
$$

Step 4. Unconstrained formulation of the problem (4.28). We show in this step that for each $\omega \in \Omega$ there exists $q \in \mathscr{H}_\alpha^{\mathfrak{s}}$ such that the constraint $\overline{u}_0(\omega) := v_\lambda[q](0, \omega) + u_0(\omega) \in \mathfrak{M}_\lambda(\omega)$ is satisfied.

Since $v_\lambda[q](\cdot, \omega)$ is the fixed point of $\mathscr{L}_q^{\omega,\lambda}$ given in (4.27), it holds that

$$q = P_{\mathfrak{s}} v_\lambda[q](0, \omega). \tag{4.37}$$

Note also that an initial datum $\overline{u}_0(\omega)$ given by $v_\lambda[q](0,\omega)+u_0(\omega)$ belongs to $\mathfrak{M}_\lambda(\omega)$ if $\overline{u}_0(\omega) = p + h_\lambda(p,\omega)$ for some $p \in \mathcal{H}^{\mathfrak{c}}$, where h_λ is the random invariant manifold function of Eq. (4.1) as provided by Theorem 4.1. Since $q = P_{\mathfrak{s}} v_\lambda[q](0,\omega)$, it is thus natural to seek for q of the following form:

$$q = P_{\mathfrak{s}} v_\lambda[q](0,\omega) = P_{\mathfrak{s}}(\overline{u}_0(\omega) - u_0(\omega)) = h_\lambda(p,\omega) - P_{\mathfrak{s}} u_0(\omega). \tag{4.38}$$

The constraint $\overline{u}_0(\omega) \in \mathfrak{M}_\lambda(\omega)$ is then written as a functional equation in terms of p:

$$\begin{aligned} p = P_{\mathfrak{c}}\overline{u}_0(\omega) &= P_{\mathfrak{c}} v_\lambda[q](0,\omega) + P_{\mathfrak{c}} u_0(\omega) \\ &= P_{\mathfrak{c}} v_\lambda[h_\lambda(p,\omega) - P_{\mathfrak{s}} u_0(\omega)](0,\omega) + P_{\mathfrak{c}} u_0(\omega). \end{aligned} \tag{4.39}$$

Note also that $P_{\mathfrak{c}} v_\lambda[h_\lambda(p,\omega) - P_{\mathfrak{s}} u_0(\omega)](0,\omega)$ has a natural integral representation. Indeed, since $v_\lambda[q](\cdot,\omega)$ is the fixed point of $\mathcal{L}_q^{\omega,\lambda}$ for each $q \in \mathcal{H}_\alpha^{\mathfrak{s}}$, we obtain by applying $P_{\mathfrak{c}}$ to (4.27) and setting t to 0 that

$$P_{\mathfrak{c}} v_\lambda[q](0,\omega) = -\int_0^{+\infty} \mathfrak{T}_{\lambda,\sigma}(0,s;\omega) P_{\mathfrak{c}} \delta G(\theta_s\omega, v_\lambda[q](s,\omega))\,\mathrm{d}s, \quad \forall\, q \in \mathcal{H}_\alpha^{\mathfrak{s}}, \tag{4.40}$$

where we used the fact that the solution operator $\mathfrak{T}_{\lambda,\sigma}$ leaves invariant the subspaces $\mathcal{H}^{\mathfrak{c}}$ and $\mathcal{H}^{\mathfrak{s}}$ as pointed out in Sect. 3.4.

It follows from (4.39) and (4.40) that the original fixed point problem (4.28), i.e., $\mathcal{L}_q^{\omega,\lambda}[v] = v$ with the constraint $v(0,\omega) + u_0(\omega) \in \mathfrak{M}_\lambda(\omega)$, can be transformed into the following unconstrained fixed point problem for p:

$$p = \mathcal{J}_{\omega,\lambda}(p), \tag{4.41}$$

where the operator $\mathcal{J}_{\omega,\lambda}\colon \mathcal{H}^{\mathfrak{c}} \to \mathcal{H}^{\mathfrak{c}}$ is defined by:

$$\begin{aligned} \mathcal{J}_{\omega,\lambda}(p) &:= P_{\mathfrak{c}} v_\lambda[q(p)](0,\omega) + P_{\mathfrak{c}} u_0(\omega) \\ &= -\int_0^{+\infty} \mathfrak{T}_{\lambda,\sigma}(0,s;\omega) P_{\mathfrak{c}} \delta G(\theta_s\omega, v_\lambda[q(p)](s,\omega))\,\mathrm{d}s + P_{\mathfrak{c}} u_0(\omega), \end{aligned} \tag{4.42}$$

with $v_\lambda[q(p)](\cdot,\omega)$ being the fixed point of $\mathcal{L}_{q(p)}^{\omega,\lambda}$ and $q(p) = h_\lambda(p,\omega) - P_{\mathfrak{s}} u_0(\omega)$.

It is then sufficient to solve (4.41) in $\mathcal{H}^{\mathfrak{c}}$ in order to solve the constrained fixed point problem (4.28) and therefore the asymptotic completeness problem; the temperedness of $\overline{u}_0$ being checked in Step 5.

We show in the following that $\mathcal{J}_{\omega,\lambda}$ is a contraction mapping on $\mathcal{H}^{\mathfrak{c}}$, leading then to a unique fixed point $p_{\omega,\lambda}(u_0)$ of $\mathcal{J}_{\omega,\lambda}$ for each given u_0.

For any given p_1 and $p_2 \in \mathcal{H}^{\mathfrak{c}}$, we get

$$\begin{aligned}\|\mathscr{J}_{\omega,\lambda}(p_1) - \mathscr{J}_{\omega,\lambda}(p_2)\|_\alpha &= \|P_{\mathfrak{c}} v_\lambda[q(p_1)](0,\omega) - P_{\mathfrak{c}} v_\lambda[q(p_2)](0,\omega)\|_\alpha \\ &= \Big\| \int_0^{+\infty} \mathfrak{T}_{\lambda,\sigma}(0,s;\omega) P_{\mathfrak{c}} \Big(\delta G\big(\theta_s\omega, v_\lambda[q(p_1)](s,\omega)\big) \\ &\quad - \delta G\big(\theta_s\omega, v_\lambda[q(p_2)](s,\omega)\big)\Big)\,\mathrm{d}s \Big\|_\alpha \\ &\le K\mathrm{Lip}(F) \int_0^\infty e^{-\eta_1 s - \int_0^s z_\sigma(\theta_\tau\omega)\,\mathrm{d}\tau} \|v_\lambda[q(p_1)](s,\omega) \\ &\quad - v_\lambda[q(p_2)](s,\omega)\|_\alpha\,\mathrm{d}s, \end{aligned} \tag{4.43}$$

where we used the partial-dichotomy estimate (3.46c) and the fact that G and F have the same Lipschitz constant $\mathrm{Lip}(F)$ to obtain the last inequality above.

Recalling that $q(p_i) = h_\lambda(p_i,\omega) - P_{\mathfrak{s}} u_0(\omega)$, we obtain the following estimates by applying (4.35) to the RHS of the last inequality above:

$$\begin{aligned} &K\mathrm{Lip}(F) \int_0^\infty e^{-\eta_1 s - \int_0^s z_\sigma(\theta_\tau\omega)\,\mathrm{d}\tau} \|v_\lambda[q(p_1)](s,\omega) - v_\lambda[q(p_2)](s,\omega)\|_\alpha\,\mathrm{d}s \\ &\le \frac{K^2\mathrm{Lip}(F)(\eta_1-\eta)^{-1}}{1-\Upsilon_1(F)} \|h_\lambda(p_1,\omega) - h_\lambda(p_2,\omega)\|_\alpha \\ &\le \frac{K^2\mathrm{Lip}(F)\mathrm{Lip}(h_\lambda)(\eta_1-\eta)^{-1}}{1-\Upsilon_1(F)} \|p_1 - p_2\|_\alpha, \end{aligned} \tag{4.44}$$

which—by noting that $K\mathrm{Lip}(F)(\eta_1-\eta)^{-1} < 1$ from (4.7)—leads to

$$\|\mathscr{J}_{\omega,\lambda}(p_1) - \mathscr{J}_{\omega,\lambda}(p_2)\|_\alpha \le \frac{K\mathrm{Lip}(h_\lambda)}{1-\Upsilon_1(F)} \|p_1 - p_2\|_\alpha, \qquad \forall\ p_1,\ p_2 \in \mathscr{H}^{\mathfrak{c}}.$$

Hence, $\mathscr{J}_{\omega,\lambda}$ is a contraction mapping on $\mathscr{H}^{\mathfrak{c}}$ by assumption (4.16).

From what precedes, given u_0 and the corresponding fixed point $p_{\omega,\lambda}(u_0)$ of the operator $\mathscr{J}_{\omega,\lambda}$, we get therefore the existence of $\overline{u}_0(\omega) = v_\lambda\big[q(p_{\omega,\lambda}(u_0))\big](0,\omega) + u_0(\omega)$ that belongs to $\mathfrak{M}_\lambda(\omega)$ so that $\overline{u}_\lambda = v_\lambda[q(p_{\omega,\lambda}(u_0))] + u_\lambda$ is a mild solution—hence also a classical solution according to Proposition A.1—of Eq. (4.1) on the manifold, and $\overline{u}_\lambda(\cdot,\omega) - u_\lambda(\cdot,\omega) \in C_\eta^+$ for each ω. We are just left with the verification that $\overline{u}_0$ is tempered.

Step 5. Temperedness of $\overline{u}_0$ and the asymptotic completeness. We saw in the previous step that given an $\mathscr{H}_\alpha$-valued tempered random initial datum u_0, the sought solution $\overline{u}_\lambda$, which solves the problem (4.28), emanates from $\overline{u}_0$ given by:

$$\overline{u}_0(\omega) = p_{\omega,\lambda}(u_0) + h_\lambda(p_{\omega,\lambda}(u_0),\omega), \tag{4.45}$$

where $p_{\omega,\lambda}(u_0)$ is the corresponding fixed point of the operator $\mathscr{J}_{\omega,\lambda}$ given in (4.42). In what follows we check that such a $\overline{u}_0$ is an $\mathscr{H}_\alpha$-valued tempered random variable.

First, we check that $\overline{u}_0$ is a random variable. Since $\overline{u}_0$ is given by $\overline{u}_0(\omega) = p_{\omega,\lambda}(u_0) + h_\lambda(p_{\omega,\lambda}(u_0), \omega)$ for each ω and h_λ is measurable, we only need to check that $\omega \mapsto p_{\omega,\lambda}(u_0)$ is $(\mathscr{F}; \mathscr{B}(\mathscr{H}^{\mathfrak{c}}))$-measurable. In that respect, note that the function $v_\lambda[q]$ involved in the definition of the operator $\mathscr{J}_{\omega,\lambda}$ is known to be $(\mathscr{B}([0,\infty)) \otimes \mathscr{F}; \mathscr{B}(\mathscr{H}_\alpha))$-measurable according to Step 3. The desired measurability of $\omega \mapsto p_{\omega,\lambda}(u_0)$ can be thus obtained by relying on the Picard scheme mentioned before associated with here the fixed point problem $\mathscr{J}_{\omega,\lambda}(p) = p$.

We check now that $\overline{u}_0$ is tempered. This is simply obtained from an appropriate control of $\|\overline{u}_0(\omega)\|_\alpha$ by $\|u_0(\omega)\|_\alpha$ which is made possible from the construction of $\overline{u}_0$ given in (4.45).

In that respect, first note that by the construction of $\mathscr{J}_{\omega,\lambda}$ given in (4.42), it holds that

$$\begin{aligned}\|p_{\omega,\lambda}(u_0)\|_\alpha &= \|\mathscr{J}_{\omega,\lambda}(p_{\omega,\lambda}(u_0))\|_\alpha \\ &\le \|P_{\mathfrak{c}} v_\lambda\big[q(p_{\omega,\lambda}(u_0))\big](0,\omega)\|_\alpha + \|P_{\mathfrak{c}} u_0(\omega)\|_\alpha \\ &\le \|P_{\mathfrak{c}} v_\lambda\big[q(p_{\omega,\lambda}(u_0))\big](\cdot,\omega)\|_{C_\eta^+} + \|P_{\mathfrak{c}} u_0(\omega)\|_\alpha.\end{aligned}$$

By applying (4.36) we obtain then for all $\omega \in \Omega$ that:

$$\begin{aligned}\|p_{\omega,\lambda}(u_0)\|_\alpha &\le \frac{K}{1-\Upsilon_1(F)}\|q(p_{\omega,\lambda}(u_0))\|_\alpha + \|P_{\mathfrak{c}} u_0(\omega)\|_\alpha \\ &= \frac{K}{1-\Upsilon_1(F)}\|h_\lambda(p_{\omega,\lambda}(u_0),\omega) - P_{\mathfrak{s}} u_0(\omega)\|_\alpha + \|P_{\mathfrak{c}} u_0(\omega)\|_\alpha \\ &\le \frac{K}{1-\Upsilon_1(F)}\big(\mathrm{Lip}(h_\lambda)\|p_{\omega,\lambda}(u_0)\|_\alpha + \|P_{\mathfrak{s}} u_0(\omega)\|_\alpha\big) + \|P_{\mathfrak{c}} u_0(\omega)\|_\alpha.\end{aligned} \tag{4.46}$$

Using assumption (4.16) in the last inequality, we obtain after simplification that

$$\begin{aligned}&\|p_{\omega,\lambda}(u_0)\|_\alpha \\ &\quad \le \frac{1-\Upsilon_1(F)}{1-\Upsilon_1(F) - K\mathrm{Lip}(h_\lambda)}\Big(\frac{K}{1-\Upsilon_1(F)}\|P_{\mathfrak{s}} u_0(\omega)\|_\alpha + \|P_{\mathfrak{c}} u_0(\omega)\|_\alpha\Big).\end{aligned} \tag{4.47}$$

By noting that $\|P_{\mathfrak{c}} u_0(\omega)\|_\alpha < \frac{K}{1-\Upsilon_1(F)}\|P_{\mathfrak{c}} u_0(\omega)\|_\alpha$ (recalling that $K \ge 1$ and $0 < \Upsilon_1(F) < 1$), we deduce:

$$\|p_{\omega,\lambda}(u_0)\|_\alpha \le \frac{2K}{1-\Upsilon_1(F) - K\mathrm{Lip}(h_\lambda)}\|u_0(\omega)\|_\alpha, \quad \forall\, \omega \in \Omega.$$

Hence, from the definition of $\overline{u}_0$ in (4.45):

$$\begin{aligned}\|\overline{u}_0(\omega)\|_\alpha &\le \|p_{\omega,\lambda}(u_0)\|_\alpha + \|h_\lambda(p_{\omega,\lambda}(u_0),\omega)\|_\alpha \\ &\le (1+\mathrm{Lip}(h_\lambda))\|p_{\omega,\lambda}(u_0)\|_\alpha \\ &\le \frac{2K(1+\mathrm{Lip}(h_\lambda))}{1-\Upsilon_1(F)-K\mathrm{Lip}(h_\lambda)}\|u_0(\omega)\|_\alpha, \quad \forall\ \omega\in\Omega.\end{aligned} \tag{4.48}$$

The temperedness property of $\overline{u}_0$ follows then from the assumption that the initial datum u_0 is tempered.

Finally, we conclude about the asymptotic completeness property. Let $\overline{u}_\lambda$ be the solution to Eq. (4.1) which solves the problem (4.28), emanating from the initial datum as given in (4.45).

From the definition of C^+_η-norm, we obtain for each $t \ge 0$ that

$$\begin{aligned}\|\overline{u}_\lambda(t,\omega)-u_\lambda(t,\omega)\|_\alpha &= \|v_\lambda[q(p_{\omega,\lambda}(u_0))](t,\omega)\|_\alpha \\ &\le e^{\eta t+\int_0^t z_\sigma(\theta_s\omega)\,\mathrm{d}s}\|v_\lambda[q(p_{\omega,\lambda}(u_0))](\cdot,\omega)\|_{C^+_\eta}.\end{aligned}$$

Using (4.36) and $q(p_{\omega,\lambda}(u_0)) = h_\lambda(p_{\omega,\lambda}(u_0),\omega) - P_{\mathfrak{s}}u_0(\omega)$ in the above inequality, we obtain

$$\|\overline{u}_\lambda(t,\omega)-u_\lambda(t,\omega)\|_\alpha \le \frac{Ke^{\eta t+\int_0^t z_\sigma(\theta_s\omega)\,\mathrm{d}s}}{1-\Upsilon_1(F)}\|h_\lambda(p_{\omega,\lambda}(u_0),\omega)-P_{\mathfrak{s}}u_0(\omega)\|_\alpha, \tag{4.49}$$

which leads to

$$\|\overline{u}_\lambda(t,\omega)-u_\lambda(t,\omega)\|_\alpha \le \frac{Ke^{\eta t+\int_0^t z_\sigma(\theta_s\omega)\,\mathrm{d}s}}{1-\Upsilon_1(F)}\|P_{\mathfrak{s}}\overline{u}_0(\omega)-P_{\mathfrak{s}}u_0(\omega)\|_\alpha, \tag{4.50}$$

by simply applying $P_{\mathfrak{s}}$ to (4.45).

For any $\varepsilon \in (0,|\eta|)$, let us now introduce the following positive random variables

$$C_{\varepsilon,\sigma}(\omega) := \sup_{t\ge 0}\frac{Ke^{-\varepsilon t+\int_0^t z_\sigma(\theta_s\omega)\,\mathrm{d}s}}{1-\Upsilon_1(F)}, \quad C'_{\varepsilon,\sigma}(\omega) := \sup_{t\ge 0}\frac{Ke^{-\varepsilon t+\int_{-t}^0 z_\sigma(\theta_s\omega)\,\mathrm{d}s}}{1-\Upsilon_1(F)}, \tag{4.51}$$

where $\Upsilon_1(F)$ is defined in (4.7).

Note that $C_{\varepsilon,\sigma}(\omega)$ and $C'_{\varepsilon,\sigma}(\omega)$ are finite for all ω because of the growth control relation of $z_\sigma(\theta_t\omega)$ given in (3.30). We obtain then from (4.50) that

$$\begin{aligned}&\|\overline{u}_\lambda(t,\omega)-u_\lambda(t,\omega)\|_\alpha \\ &\quad\le C_{\varepsilon,\sigma}(\omega)e^{(\eta+\varepsilon)t}\|P_{\mathfrak{s}}\overline{u}_0(\omega)-P_{\mathfrak{s}}u_0(\omega)\|_\alpha, \quad \forall\, t\ge 0,\ \omega\in\Omega.\end{aligned} \tag{4.52}$$

By a simple change of fiber, we also obtain from (4.50) that:

$$\begin{aligned}&\|\overline{u}_\lambda(t,\theta_{-t}\omega) - u_\lambda(t,\theta_{-t}\omega)\|_\alpha \\ &\quad \le C'_{\varepsilon,\sigma}(\omega)e^{(\eta+\varepsilon)t}\|P_{\mathfrak{s}}\overline{u}_0(\theta_{-t}\omega) - P_{\mathfrak{s}}u_0(\theta_{-t}\omega)\|_\alpha, \quad \forall\, t \ge 0,\ \omega \in \Omega.\end{aligned} \tag{4.53}$$

Recalling that $\eta < 0$, the forward and pullback asymptotic completeness of $\mathfrak{M}_\lambda$ follow respectively from (4.52) and (4.53) with $\overline{u}_0(\omega)$ defined in (4.45). In both (4.52) and (4.53), the rate of attraction is given by $|\eta| - \varepsilon$ for any $\varepsilon > 0$ so that the critical attraction rate is $|\eta|$. The proof is now complete. □

Remark 4.4

(1) The proof of Corollary 4.3 follows from the cohomology relation (3.41) and the estimate (4.50). Indeed, the same estimates as given in (4.52) and (4.53) for the SPDE case can be derived from (4.50) using the cohomology relation, but with the following random coefficients $\widehat{C}_{\varepsilon,\sigma}(\omega)$ and $\widehat{C}'_{\varepsilon,\sigma}(\omega)$:

$$\widehat{C}_{\varepsilon,\sigma}(\omega) := \sup_{t\ge 0} \frac{K e^{-\varepsilon t + \sigma W_t(\omega)}}{1 - \Upsilon_1(F)}, \quad \widehat{C}'_{\varepsilon,\sigma}(\omega) := \sup_{t\ge 0} \frac{K e^{-\varepsilon t - \sigma W_{-t}(\omega)}}{1 - \Upsilon_1(F)}. \tag{4.54}$$

(2) Note that the random positive variable $\widehat{C}_{\varepsilon,\sigma}$ exhibits fluctuations which get larger as σ gets bigger. In particular, given $\gamma > 0$, if we denote by $t^*_\sigma(\omega)$ the minimal time after which $\|\overline{u}_\lambda(t,\omega) - u_\lambda(t,\omega)\|_\alpha$ gets smaller than γ, it can be shown that its expected value m_σ increases with σ, while as σ tends to zero, m_σ converges to the corresponding attraction time associated with $\sigma = 0$. Hence, although the critical rate of attraction is independent of σ, the expected attraction time to the inertial manifold (for a given precision γ) is not.
(3) Note that from the Proof of Theorem 4.3, we see that in the case where $\eta > 0$ the difference between the given solution u_λ of Eq. (4.1) and the constructed solution $\overline{u}_\lambda$ living on the random invariant manifold $\mathfrak{M}_\lambda$ has a growth controlled by $C_{\varepsilon,\sigma}(\omega)e^{(\eta+\varepsilon)t}$ in the forward case (resp. $C'_{\varepsilon,\sigma}(\omega)e^{(\eta+\varepsilon)t}$ in the pullback case) for any $\varepsilon > 0$.

Chapter 5
Local Stochastic Invariant Manifolds: Preparation to Critical Manifolds

In this chapter we present a local theory of stochastic invariant manifolds associated with the global theory described in Chap. 4. The ideas are standard but the precise framework is detailed here again in view of the main results regarding the approximation formulas of stochastic critical manifolds (Chap. 6) and the related pullback characterizations presented in Volume II [37, Sect. 4.1]. In particular, the proof of Theorem 5.1 is provided below where some elements will be used in establishing Theorem 6.1.

We introduce now the following notion of a local random (resp. stochastic) invariant manifold on which we will rely to build our theory of stochastic critical manifolds in Chap. 6.

Definition 5.1 Let S be a continuous RDS acting on some separable Hilbert space H, and r be a given positive integer. A random closed set $\mathfrak{M}^{\text{loc}}$ is called a local random invariant C^r-manifold of S if the following conditions hold:

(i) $\mathfrak{M}^{\text{loc}}$ is locally invariant in the sense that for each ω and each $u_0 \in \mathring{\mathfrak{M}}^{\text{loc}}(\omega)$ ($\neq \emptyset$),[1] there exists $t_{u_0,\omega} > 0$, such that $S(t,\omega)u_0 \in \mathring{\mathfrak{M}}^{\text{loc}}(\theta_t\omega)$ for all $t \in [0, t_{u_0,\omega})$.
(ii) There exist a closed subspace $\mathscr{G} \subset H$ and a measurable function $h\colon \mathscr{G} \times \Omega \to \mathscr{L}$, with $\mathscr{L}$ the topological complement of $\mathscr{G}$ in H, such that $\mathfrak{M}^{\text{loc}}(\omega)$ can be represented as the graph of $h(\cdot,\omega)|_{\mathfrak{B}(\omega)}$ for each ω, i.e.,

$$\mathfrak{M}^{\text{loc}}(\omega) = \{\xi + h(\xi,\omega) \mid \xi \in \mathfrak{B}(\omega)\},$$

where $\mathfrak{B}\colon \Omega \to 2^{\mathscr{G}}$ is some random closed ball, such that $\mathfrak{B}(\omega)$ is centered at zero for each ω.
(iii) $h(\cdot,\omega)|_{\mathfrak{B}(\omega)}$ is C^r for all ω.
(iv) $\mathfrak{M}^{\text{loc}}(\omega)$ is tangent to the subspace $\mathscr{G}$ at the origin:

$$h(0,\omega) = 0, \quad D_\xi h(\xi,\omega)\big|_{\xi=0} = 0, \quad \forall\, \omega \in \Omega.$$

[1]Here, $\mathring{\mathfrak{M}}^{\text{loc}}(\omega)$ denotes the interior of $\mathfrak{M}^{\text{loc}}(\omega)$.

M.D. Chekroun et al., *Approximation of Stochastic Invariant Manifolds*,
SpringerBriefs in Mathematics, DOI 10.1007/978-3-319-12496-4_5

The function h is called the local random invariant manifold function associated with $\mathfrak{M}^{\text{loc}}$.

As in the global version, we will refer the associated local manifold as a local stochastic invariant C^r-manifold when we will deal with Eq. (3.1).

Remark 5.1.

(1) The random closed ball $\mathfrak{B}$ in part (ii) of the above definition can be deterministic; see, e.g., Corollary 5.1 for such a situation.
(2) Note that this notion of local random invariant manifold is a natural generalization of the classical one already encountered for non-autonomous dynamical systems; see, e.g., [92, Definition 6.1.1]. Here, "local" has to be understood in terms of both the time variable t and the phase-space variable ξ as specified in (i) and (ii) in the definition above. As we will see in Step 3 of the proof of Theorem 5.1, this "local" property arises as a direct consequence of the cut-off procedure.
(3) Note also that Definition 5.1 does not exclude that

$$S(t,\omega)u_0 \in \mathring{\mathfrak{M}}^{\text{loc}}(\theta_t\omega),$$

for $t \in [t^1_{u_0,\omega}, t^2_{u_0,\omega}]$ with $t^2_{u_0,\omega} > t^1_{u_0,\omega} > t_{u_0,\omega}$. Actually, the noise effects can make this to happen on infinitely-many such time intervals.

Let us introduce now a cut-off version of the nonlinearity F. Let $\zeta : \mathbb{R}^+ \to [0,1]$ be a C^∞ decreasing function, such that

$$\zeta(s) = \begin{cases} 1 & \text{if } s \leq 1, \\ 0 & \text{if } s \geq 2. \end{cases} \tag{5.1}$$

For any given positive constant ρ, let us define a mapping $F_\rho : \mathcal{H}_\alpha \to \mathcal{H}$ via

$$F_\rho(u) := \zeta\left(\frac{\|u\|_\alpha}{\rho}\right)F(u). \tag{5.2}$$

We are now in position to formulate the results regarding the existence and smoothness of families of local random invariant manifolds for Eq. (4.1), when the latter generates a continuous (global) RDS acting on $\mathcal{H}_\alpha$.[2]

Theorem 5.1 *Consider the RPDE* (4.1). *The linear operator L_λ is assumed to satisfy the corresponding assumptions in Sect.* 3.1. *The nonlinearity $F : \mathcal{H}_\alpha \to \mathcal{H}$ is*

[2]For applications we have in mind, it is often met that for any given initial datum $u_0 \in \mathcal{H}_\alpha$ there exists a unique classical solution to Eq. (4.1) for all $t \geq 0$ in the sense given in Proposition 3.1, and hence Eq. (4.1) generates an RDS in the sense given in Sect. 3.2. Such property holds for a broad class of random dissipative evolution equations; see, e.g., [55, 85, 106].

assumed to be C^p*-smooth for some integer* $p \geq 2$*, and to satisfy* (3.5) *and* (3.6).[3] *Assume also that Eq.* (4.1) *generates a continuous RDS acting on the space* $\mathcal{H}_\alpha$.

We assume furthermore that an open interval Λ *is chosen such that the uniform spectrum decomposition* (3.11) *holds over* Λ*, and such that there exist* η_1 *and* η_2 *satisfying*

$$\eta_{\mathfrak{c}} > \eta_1 > \eta_2 > \eta_{\mathfrak{s}}, \tag{5.3}$$

for which

$$\exists\, \eta \in (\eta_2, \eta_1) \text{ s.t. } \eta_2 < j\eta < \eta_1, \quad \forall\, j \in \{1, \dots, r\}, \tag{5.4}$$

for some integer $r \in \{1, \dots, p\}$. *As above,* $\eta_{\mathfrak{c}}$ *and* $\eta_{\mathfrak{s}}$ *are defined in* (3.12)*, and* $\mathcal{H}^{\mathfrak{c}}$ *and* $\mathcal{H}_\alpha^{\mathfrak{s}}$ *denote the subspaces associated with the uniform spectrum decomposition, defined in* (3.18) *and* (3.20) *with* $\dim(\mathcal{H}^{\mathfrak{c}}) = m$.

Under these conditions, there exists $\rho^* > 0$ *such that for each* $\rho \in (0, \rho^*)$ *the following uniform spectral gap conditions hold:*

$$\boxed{\Upsilon_1(F_\rho) = K Lip(F_\rho)\big((\eta_1 - \eta)^{-1} + \Gamma(1-\alpha)(\eta - \eta_2)^{\alpha-1}\big) < \frac{1}{2},} \tag{5.5}$$

and

$$\boxed{\Upsilon_j(F_\rho) = K Lip(F_\rho)\big((\eta_1 - j\eta)^{-1} + \Gamma(1-\alpha)(j\eta - \eta_2)^{\alpha-1}\big) < 1, \ \forall\, j \in \{2, \dots, r\},} \tag{5.6}$$

where F_ρ *is defined by* (5.2) *and* K *is as given in the partial-dichotomy estimates* (3.24).

Moreover, for each $\rho \in (0, \rho^*)$*, Eq.* (4.1) *possesses a family of local random invariant* C^r*-manifolds* $\{\mathfrak{M}_\lambda^{\mathrm{loc}}\}_{\lambda \in \Lambda}$*, where each of such manifolds is m-dimensional and is given by*

$$\mathfrak{M}_\lambda^{\mathrm{loc}}(\omega) := \{\xi + h_\lambda(\xi, \omega) \mid \xi \in \mathfrak{B}_\rho(\omega)\}, \quad \omega \in \Omega,\ \lambda \in \Lambda, \tag{5.7}$$

with $h_\lambda \colon \mathcal{H}^{\mathfrak{c}} \times \Omega \to \mathcal{H}_\alpha^{\mathfrak{s}}$ *denoting the corresponding random invariant manifold function. The associated random closed ball* $\mathfrak{B}_\rho$ *is given by*[4]:

$$\mathfrak{B}_\rho(\omega) := \overline{B_\alpha(0, \delta_\rho(\omega))} \cap \mathcal{H}^{\mathfrak{c}}, \ \omega \in \Omega, \tag{5.8}$$

with

$$\delta_\rho(\omega) := \frac{e^{-z_\sigma(\omega)}\rho}{1+K}, \quad \omega \in \Omega, \tag{5.9}$$

[3] But not necessarily required to be globally Lipschitz here.

[4] Throughout this monograph, for any given positive random variable $r(\omega)$, $B_\alpha(0, r)$ denotes the random open ball in $\mathcal{H}_\alpha$ centered at the origin with random radius r; see, e.g., [54, Definition 2.1].

where $z_\sigma(\omega)$ is the OU process defined in (3.29).

Thanks to the cohomology relation between the two RDSs associated with Eqs. (3.1) and (4.1) as recalled in Sect. 3.3, we deduce the following corollary; the proof of Theorem 5.1 being provided below.

Corollary 5.1 *Consider the SPDE* (3.1). *Assume that the same assumptions as in Theorem* 5.1 *hold with Λ specified therein. Let also ρ^* be the positive constant as in Theorem* 5.1.

Then for each $\rho \in (0, \rho^*)$, Eq. (3.1) *possesses a family of local stochastic invariant C^r-manifolds $\{\widehat{\mathfrak{M}}_\lambda^{\mathrm{loc}}\}_{\lambda\in\Lambda}$, where each of such manifolds is m-dimensional and is given by*

$$\widehat{\mathfrak{M}}_\lambda^{\mathrm{loc}}(\omega) := \{\xi + \widehat{h}_\lambda(\xi, \omega) \mid \xi \in \widehat{\mathfrak{B}}_\rho\}, \quad \omega \in \Omega, \ \lambda \in \Lambda, \tag{5.10}$$

with $\widehat{h}_\lambda : \mathscr{H}^{\mathfrak{c}} \times \Omega \to \mathscr{H}_\alpha^{\mathfrak{s}}$ denoting the corresponding stochastic invariant manifold function for each $\lambda \in \Lambda$.

In this case, the associated "random closed ball" is simply a deterministic ball, $\widehat{\mathfrak{B}}_\rho$, which is given by

$$\widehat{\mathfrak{B}}_\rho := \overline{B_\alpha(0, \widehat{\delta}_\rho)} \cap \mathscr{H}^{\mathfrak{c}}, \tag{5.11}$$

with

$$\widehat{\delta}_\rho := \frac{\rho}{(1+K)}. \tag{5.12}$$

Moreover, the following relation holds:

$$\widehat{h}_\lambda(\xi, \omega) = e^{z_\sigma(\omega)} h_\lambda(e^{-z_\sigma(\omega)}\xi, \omega), \quad \xi \in \mathscr{H}^{\mathfrak{c}}, \ \omega \in \Omega, \tag{5.13}$$

where $h_\lambda(\xi, \omega)$ is specified in Theorem 5.1.

Remark 5.2.

(1) Note that from the proof of Theorem 5.1 given below (see, e.g., (5.39)), for each $\rho \in (0, \rho^*)$, the local random invariant manifold $\mathfrak{M}_\lambda^{\mathrm{loc}}$ can actually be obtained as the random graph over a larger random closed ball with random radius given by

$$\widetilde{\delta}_\rho(\omega) := \frac{e^{-z_\sigma(\omega)}\rho}{1 + \mathrm{Lip}(h_{\lambda,\rho})}, \tag{5.14}$$

since indeed $\delta_\rho(\omega) < \widetilde{\delta}_\rho(\omega)$ for all ω according to (5.41).
Similarly, the local stochastic invariant manifold $\widehat{\mathfrak{M}}_\lambda^{\mathrm{loc}}$ given in Corollary 5.1 can also be defined over a larger deterministic ball with radius given by $\frac{\rho}{1+\mathrm{Lip}(h_{\lambda,\rho})}$.
In all the cases, it is important to note that such deterministic radii are artificial and result from the techniques employed in the proofs of Theorem 5.1 and Corollary 5.1, such as, e.g., the cohomology approach; see also Remark 6.4.

(2) The radius δ_ρ provided in Theorem 5.1 results from a control of $\mathrm{Lip}(h_{\lambda,\rho})$ by K made possible by the choice of ρ^*; see Step 3 below. The interest is that the resulting manifolds $\mathfrak{M}_\lambda^{\mathrm{loc}}$ live then above some λ-independent neighborhoods, which help simplify certain algebraic manipulation in the proof of Theorem 6.1 that relies on Theorem 5.1.

Proof of Theorem 5.1 *As mentioned before, the proof is accomplished via a cut-off procedure. We proceed in three steps.*

Step 1. Modification of the nonlinearity F through a cut-off function. We show in this first step that the modified nonlinearity F_ρ defined in (5.2) is globally Lipschitz while still being C^p-smooth.

We first examine the smoothness of F_ρ. Note that the $\|\cdot\|_\alpha$-norm is C^∞-smooth at any nonzero element in $\mathcal{H}_\alpha$ as norm induced by the inner product $\langle\cdot,\cdot\rangle_\alpha$ given in Chap. 3. Then, a basic composition argument leads to C^p smoothness of F_ρ at any nonzero element. It is also clear that F_ρ has this smoothness at the origin by the construction of ζ; and F_ρ is thus C^p-smooth on $\mathcal{H}_\alpha$.

Now, we show that F_ρ is globally Lipschitz when ρ is sufficiently small. Note that DF is continuous since F is C^p-smooth with $p \geq 2$ by our assumption. Note also that $DF(0) = 0$ by (3.6). Then, there exists $\rho_1 > 0$ and a positive function $C(\rho)$ defined on $[0, \rho_1]$ such that

$$C(\rho) \to 0 \quad \text{as} \quad \rho \to 0, \tag{5.15}$$

and

$$\|DF(u)\|_{\alpha,0} \leq C(\rho), \quad \forall\ \rho \in (0, \rho_1],\ u \in B_\alpha(0,\rho) \subset \mathcal{H}_\alpha, \tag{5.16}$$

where $\|DF(u)\|_{\alpha,0}$ stands for the operator norm of $DF(u)$ as a linear operator from $\mathcal{H}_\alpha$ to $\mathcal{H}$.

For any $u_1, u_2 \in \mathcal{H}_\alpha$, by the mean value theorem (see, e.g., [112, Theorem 4.2]), we obtain

$$\begin{aligned} F(u_2) - F(u_1) &= \int_0^1 DF((1-s)u_1 + su_2)(u_2 - u_1)\,\mathrm{d}s \\ &= \int_0^1 DF((1-s)u_1 + su_2)\,\mathrm{d}s\,(u_2 - u_1). \end{aligned} \tag{5.17}$$

Now, it follows from (5.16) and (5.17) that

$$\|F(u_1) - F(u_2)\| \leq \left\| \int_0^1 DF((1-s)u_1 + su_2)\,\mathrm{d}s \right\|_{\alpha,0} \|(u_2 - u_1)\|_\alpha$$

$$\le \Big(\int_0^1 \|DF((1-s)u_1 + su_2)\|_{\alpha,0}\, \mathrm{d}s\Big) \|(u_2 - u_1)\|_\alpha \tag{5.18}$$
$$\le C(\rho)\|u_1 - u_2\|_\alpha, \quad \forall\ \rho \in (0, \rho_1],\ u_1, u_2 \in B_\alpha(0, \rho).$$

Recall that $D^2 F$ is also continuous. In particular, it is continuous at the origin. There exists then $\rho_2 > 0$ and a bounded positive function $\widetilde{C}(\rho)$ defined on $[0, \rho_2]$ such that

$$\|D^2 F(u)\|_{\mathrm{op}} \le \widetilde{C}(\rho), \quad \forall\ \rho \in (0, \rho_2],\ u \in B_\alpha(0, \rho), \tag{5.19}$$

where $\|D^2 F(u)\|_{\mathrm{op}}$ is the operator norm of $D^2 F(u)$ as a bilinear operator from $\mathscr{H}_\alpha \times \mathscr{H}_\alpha$ to $\mathscr{H}$.

Again, since F is C^p-smooth with $p \ge 2$, $F(0) = 0$, and $DF(0) = 0$, it follows then from the Taylor-Lagrange theorem (cf. [112, p. 349]) that

$$F(u) = \int_0^1 (1-s) D^2 F(su)(u, u)\, \mathrm{d}s, \quad \forall\, u \in \mathscr{H}_\alpha. \tag{5.20}$$

We obtain from (5.19) and (5.20) that

$$\|F(u)\| \le \widetilde{C}(\rho)\|u\|_\alpha^2, \quad \forall\ \rho \in (0, \rho_2],\ u \in B_\alpha(0, \rho). \tag{5.21}$$

Now, let

$$\rho_3 := \min\left\{\frac{\rho_1}{2}, \frac{\rho_2}{2}\right\}. \tag{5.22}$$

We show that for each fixed $\rho \in (0, \rho_3)$ it holds that

$$\begin{aligned} &\|F_\rho(u_1) - F_\rho(u_2)\| \\ &\qquad \le \big(C(2\rho) + 4\rho\widetilde{C}(2\rho)\mathrm{Lip}(\zeta)\big)\|u_1 - u_2\|_\alpha, \ \ \forall\, u_1, u_2 \in \mathscr{H}_\alpha, \end{aligned} \tag{5.23}$$

where $\mathrm{Lip}(\zeta)$ denotes the global Lipschitz constant of the cut-off function ζ:

$$|\zeta(s_1) - \zeta(s_2)| \le \mathrm{Lip}(\zeta)|s_1 - s_2|, \quad \forall\, s_1, s_2 \in \mathbb{R}^+. \tag{5.24}$$

There are three cases to be considered. If $\|u_1\|_\alpha, \|u_2\|_\alpha \ge 2\rho$, then (5.23) holds trivially by the definition of F_ρ and the construction of ζ.

If $\|u_1\|_\alpha < 2\rho$ and $\|u_2\|_\alpha \ge 2\rho$, we get

$$\begin{aligned} \|F_\rho(u_1) - F_\rho(u_2)\| = \|F_\rho(u_1)\| &= \Big\|\zeta\Big(\frac{\|u_1\|_\alpha}{\rho}\Big)F(u_1)\Big\| \\ &\le \widetilde{C}(2\rho)\|u_1\|_\alpha^2 \Big\|\zeta\Big(\frac{\|u_1\|_\alpha}{\rho}\Big)\Big\| \end{aligned}$$

$$
\begin{aligned}
&\leq 4\rho^2\widetilde{C}(2\rho)\Big\|\zeta\Big(\frac{\|u_1\|_\alpha}{\rho}\Big)\Big\| \\
&= 4\rho^2\widetilde{C}(2\rho)\Big\|\zeta\Big(\frac{\|u_1\|_\alpha}{\rho}\Big)-\zeta\Big(\frac{\|u_2\|_\alpha}{\rho}\Big)\Big\| \\
&\leq 4\rho\widetilde{C}(2\rho)\,\mathrm{Lip}\,(\zeta)(\|u_2\|_\alpha-\|u_1\|_\alpha) \\
&\leq 4\rho\widetilde{C}(2\rho)\,\mathrm{Lip}\,(\zeta)\|u_1-u_2\|_\alpha,
\end{aligned} \tag{5.25}
$$

where we used (5.21) and $2\rho < 2\rho_3 \leq \rho_2$ to derive the first inequality above. So (5.23) holds in this case. By exchanging the role of u_1 and u_2, the result also holds for the case when $\|u_2\|_\alpha < 2\rho$ and $\|u_1\|_\alpha \geq 2\rho$.

Finally, if $\|u_1\|_\alpha < 2\rho$ and $\|u_2\|_\alpha < 2\rho$, we get

$$
\begin{aligned}
\|F_\rho(u_1)-F_\rho(u_2)\| &= \Big\|\zeta\Big(\frac{\|u_1\|_\alpha}{\rho}\Big)F(u_1)-\zeta\Big(\frac{\|u_2\|_\alpha}{\rho}\Big)F(u_2)\Big\| \\
&\leq \Big\|\zeta\Big(\frac{\|u_1\|_\alpha}{\rho}\Big)F(u_1)-\zeta\Big(\frac{\|u_1\|_\alpha}{\rho}\Big)F(u_2)\Big\| \\
&\quad+\Big\|\zeta\Big(\frac{\|u_1\|_\alpha}{\rho}\Big)F(u_2)-\zeta\Big(\frac{\|u_2\|_\alpha}{\rho}\Big)F(u_2)\Big\| \\
&\leq C(2\rho)\|u_1-u_2\|_\alpha+\mathrm{Lip}\,(\zeta)\Big|\frac{\|u_1\|_\alpha}{\rho}-\frac{\|u_2\|_\alpha}{\rho}\Big|\cdot\|F(u_2)\| \\
&\leq C(2\rho)\|u_1-u_2\|_\alpha+4\rho\widetilde{C}(2\rho)\,\mathrm{Lip}\,(\zeta)\|u_1-u_2\|_\alpha,
\end{aligned} \tag{5.26}
$$

where we used (5.24) and (5.18) to derive the second last inequality above, and used (5.21) and $\|u_2\|_\alpha < 2\rho$ to obtain the last inequality above. Again, (5.23) follows. Thus, F_ρ is indeed globally Lipschitz for each fixed $\rho \in (0, \rho_3)$.

Now, let

$$
\mathrm{Lip}(F_\rho) := C(2\rho) + 4\rho\widetilde{C}(2\rho)\,\mathrm{Lip}\,(\zeta), \quad \rho \in (0, \rho_3). \tag{5.27}
$$

Then, it follows from (5.23) that

$$
\|F_\rho(u_1)-F_\rho(u_2)\| \leq \mathrm{Lip}(F_\rho)\|u_1-u_2\|_\alpha, \ \forall\, u_1, u_2 \in \mathscr{H}_\alpha \text{ and } \rho \in (0, \rho_3). \tag{5.28}
$$

Step 2. Existence of global random invariant manifolds for the modified equation. Now, for each fixed $\rho > 0$, let us consider the following cut-off version of Eq. (4.1):

$$
\frac{\mathrm{d}u}{\mathrm{d}t} = L_\lambda u + z_\sigma(\theta_t\omega)u + G_\rho(\theta_t\omega, u), \tag{5.29}
$$

where

$$
G_\rho(\omega, u) := e^{-z_\sigma(\omega)}F_\rho(e^{z_\sigma(\omega)}u) = e^{-z_\sigma(\omega)}\zeta\Big(\frac{e^{z_\sigma(\omega)}\|u\|_\alpha}{\rho}\Big)F(e^{z_\sigma(\omega)}u). \tag{5.30}
$$

We first check condition (4.12) for Eq. (5.29) when ρ is sufficiently small. Note that the Lipschitz constant of F_ρ given in (5.27) satisfies that

$$\mathrm{Lip}(F_\rho) \to 0 \quad \text{as} \quad \rho \to 0, \tag{5.31}$$

which follows from (5.15) and the boundedness of $\widetilde{C}(\rho)$ for sufficiently small ρ.

For η given in (5.4), we infer from (5.31) that there exists a positive constant ρ_4 such that for all $\rho \in (0, \rho_4)$, the following conditions are satisfied:

$$\Upsilon_j(F_\rho) = K\mathrm{Lip}(F_\rho)\big((\eta_1 - j\eta)^{-1} + \Gamma(1-\alpha)(j\eta - \eta_2)^{\alpha-1}\big) < 1, \ \ \forall j \in \{1, \dots, r\}, \tag{5.32}$$

where $r \in \{1, \dots, p\}$ is as given in (5.4). Condition (4.12) is thus verified with F_ρ in place of F.

It is clear that Eq. (5.29) also satisfies other assumptions required in Theorems 4.1 and 4.2. Hence, for each $\rho \in (0, \rho_4)$, there exists a family of global random invariant C^r-manifolds $\{\mathfrak{M}_{\lambda,\rho}\}_{\lambda\in\Lambda}$ for Eq. (5.29), which is given by:

$$\mathfrak{M}_{\lambda,\rho}(\omega) = \{\xi + h_{\lambda,\rho}(\xi, \omega) \mid \xi \in \mathcal{H}^{\mathfrak{c}}\}, \quad \forall\, \omega \in \Omega,\ \lambda \in \Lambda,\ \rho \in (0, \rho_4), \tag{5.33}$$

where the random invariant manifold function $h_{\lambda,\rho}\colon \mathcal{H}^{\mathfrak{c}} \times \Omega \to \mathcal{H}^{\mathfrak{s}}_\alpha$ is C^r-smooth and globally Lipschitz with respect to ξ for each ω, and it is tangent to $\mathcal{H}^{\mathfrak{c}}$ at the origin:

$$h_{\lambda,\rho}(0, \omega) = 0, \quad D_\xi h_{\lambda,\rho}(0, \omega) = 0, \quad \forall\, \omega \in \Omega. \tag{5.34}$$

Moreover, by the Lipschitz estimate given in (4.9), we also have that

$$\mathrm{Lip}(h_{\lambda,\rho}) \le \frac{K^2\mathrm{Lip}(F_\rho)(\eta - \eta_2)^{\alpha-1}\Gamma(1-\alpha)}{1 - \Upsilon_1(F_\rho)}, \quad \rho \in (0, \rho_4), \tag{5.35}$$

where $\Upsilon_1(F_\rho)$ is given in (5.5).

Step 3. Existence of local random invariant manifolds for Eq. (4.1). Now, we show that there exists $\rho^* > 0$ sufficiently small such that for any given $\rho \in (0, \rho^*)$, we can find a random closed ball centered at the origin in $\mathcal{H}^{\mathfrak{c}}$ such that for each $\lambda \in \Lambda$ the random graph of $h_{\lambda,\rho}$ over this random closed ball is a local random invariant C^r-manifold for Eq. (4.1). The construction of such manifolds relies on the obvious fact that a solution of the original Eq. (4.1) coincides with a solution of the cut-off version Eq. (5.29) over some time interval $[0, t]$ provided that they both emanate from the same sufficiently small initial datum.

We first remark that for each $\rho > 0$ by comparing G_ρ given in (5.30) with G given in (4.2), we have that

$$\forall\, u \in \mathcal{H}_\alpha, \quad \Big(\|u\|_\alpha \le e^{-z_\sigma(\omega)}\rho\Big) \Longrightarrow \Big(G_\rho(\omega, u) = G(\omega, u)\Big). \tag{5.36}$$

Note also that by the continuous dependence of the solutions of Eq. (5.29) with respect to the initial data and the continuity of the OU process $t \mapsto z_\sigma(\theta_t\omega)$, we have:

$$\forall\ u_0 \in B_\alpha(0, e^{-z_\sigma(\omega)}\rho),\ \exists\ t^*_{u_0,\omega} > 0, \quad \text{s.t.} \\ \|u_{\lambda,\rho}(t,\omega;u_0)\|_\alpha < e^{-z_\sigma(\theta_t\omega)}\rho, \quad \forall\ t \in [0, t^*_{u_0,\omega}). \tag{5.37}$$

For each $\omega \in \Omega$ and $u_0 \in \mathscr{H}_\alpha$, let us denote the solution to Eq. (4.1) emanating from u_0 (in fiber ω) by $u_\lambda(t,\omega;u_0)$, which is guaranteed to exist since Eq. (4.1) is assumed to generate an RDS. From (5.36) and (5.37) we then obtain that

$$\forall\ u_0 \in B_\alpha(0, e^{-z_\sigma(\omega)}\rho),\ \exists\ t^*_{u_0,\omega} > 0, \quad \text{s.t.} \\ u_\lambda(t,\omega;u_0) = u_{\lambda,\rho}(t,\omega;u_0), \quad t \in [0, t^*_{u_0,\omega}). \tag{5.38}$$

The idea is then to build the sought local random invariant manifold for Eq. (4.1) as a subset $\mathfrak{M}^{\text{loc}}_\lambda$ that lives in $\overline{B_\alpha(0, e^{-z_\sigma(\omega)}\rho)}$ of the (global) invariant manifold $\mathfrak{M}_{\lambda,\rho}$ for the modified equation (5.29). By taking indeed any initial datum u_0 in the interior of such a subset, it holds $\|u_0\|_\alpha < e^{-z_\sigma(\omega)}\rho$, so that $u_\lambda = u_{\lambda,\rho}$ on $[0, t^*_{u_0,\omega})$ by (5.38). As we will see, this will imply the local invariance property of $\mathfrak{M}^{\text{loc}}_\lambda$ from the invariance property of $\mathfrak{M}_{\lambda,\rho}$ under $u_{\lambda,\rho}$.

We focus then first on how to ensure $\|u_0\|_\alpha \le e^{-z_\sigma(\omega)}\rho$ when $u_0 \in \mathfrak{M}_{\lambda,\rho}$. For each such u_0, and each $\rho \in (0, \rho_4)$, we have that

$$u_0 = P_{\mathfrak{c}} u_0 + h_{\lambda,\rho}(P_{\mathfrak{c}} u_0, \omega),$$

which implies

$$\|u_0\|_\alpha \le \|P_{\mathfrak{c}} u_0\|_\alpha + \|h_{\lambda,\rho}(P_{\mathfrak{c}} u_0,\omega)\|_\alpha \le (1 + \text{Lip}(h_{\lambda,\rho}))\|P_{\mathfrak{c}} u_0\|_\alpha. \tag{5.39}$$

Recalling that $\text{Lip}(F_\rho) \to 0$ as $\rho \to 0$, we have

$$\exists\ \rho_5 > 0, \quad \text{s.t.}\ \ \Upsilon_1(F_\rho) \le \frac{1}{2}, \quad \rho \in (0, \rho_5]. \tag{5.40}$$

Therefore, by noting that $K\text{Lip}(F_\rho)(\eta - \eta_2)^{\alpha-1}\Gamma(1-\alpha) \le \Upsilon_1(F_\rho)$ from (5.5), we get from (5.35) that

$$\text{Lip}(h_{\lambda,\rho}) \le K, \qquad \forall\ \rho \in (0, \rho^*), \tag{5.41}$$

where

$$\rho^* := \min\{\rho_4, \rho_5\}, \tag{5.42}$$

leading in turn to

$$\|u_0\|_\alpha \le (1+K)\|P_{\mathfrak{c}} u_0\|_\alpha, \quad \forall\ \rho \in (0, \rho^*). \tag{5.43}$$

It is then sufficient that $\|P_c u_0\|_\alpha < \frac{e^{-z_\sigma(\omega)}\rho}{1+K}$ to ensure $\|u_0\|_\alpha \le e^{-z_\sigma(\omega)}\rho$. Let us now introduce

$$\mathfrak{B}_\rho(\omega) := \overline{B_\alpha(0, \delta_\rho(\omega))} \cap \mathscr{H}^c, \quad \rho \in (0, \rho^*), \ \omega \in \Omega, \tag{5.44}$$

with

$$\delta_\rho(\omega) := \frac{e^{-z_\sigma(\omega)}\rho}{1+K}, \quad \omega \in \Omega. \tag{5.45}$$

We claim that for each $\rho \in (0, \rho^*)$, $\mathfrak{M}_\lambda^{\mathrm{loc}}$ defined by

$$\mathfrak{M}_\lambda^{\mathrm{loc}}(\omega) := \{\xi + h_\lambda(\xi, \omega) \mid \xi \in \mathfrak{B}_\rho(\omega)\}, \quad \omega \in \Omega, \ \lambda \in \Lambda, \tag{5.46}$$

corresponds to a local random invariant C^r-manifold associated with Eq. (4.1), where the corresponding local random invariant manifold function h_λ is taken to be $h_{\lambda,\rho}$ as obtained in Step 2.

The fact that $\mathfrak{M}_\lambda^{\mathrm{loc}}$ is a random closed set can be obtained by considering the sequence $\{\gamma_n\}_{n\in\mathbb{N}}$ of measurable mappings $\gamma_n \colon \Omega \to \mathscr{H}_\alpha$ defined by $\gamma_n(\omega) := e^{-z_\sigma(\omega)}u_n + h_\lambda(e^{-z_\sigma(\omega)}u_n, \omega)$, where $\{u_n\} \in (\mathscr{H}^c)^{\mathbb{N}}$ is a sequence that is dense in $B_\alpha(0, \frac{\rho}{1+K}) \cap \mathscr{H}^c$. Then similar to Step 3 of Appendix B, we get that $\mathfrak{M}_\lambda^{\mathrm{loc}}(\omega) = \overline{\{\gamma_n(\omega) | n \in \mathbb{N}\}}^{\mathscr{H}_\alpha}$, which implies that $\mathfrak{M}_\lambda^{\mathrm{loc}}$ is a random closed set by application of the Selection Theorem [32, Theorem III.9].

Now, we show that $\mathfrak{M}_\lambda^{\mathrm{loc}}$ is locally invariant under the RDS associated with Eq. (4.1). Note that $\|P_c u_0\|_\alpha < \delta_\rho(\omega)$ for all u_0 in the interior, $\mathring{\mathfrak{M}}_\lambda^{\mathrm{loc}}(\omega)$, of $\mathfrak{M}_\lambda^{\mathrm{loc}}(\omega)$, which implies that $\|u_0\|_\alpha < e^{-z_\sigma(\omega)}\rho$ thanks to (5.43) and (5.45). Then, it follows from (5.38) that there exists $t^*_{u_0,\omega} > 0$ such that

$$u_\lambda(t, \omega; u_0) = u_{\lambda,\rho}(t, \omega; u_0), \quad t \in [0, t^*_{u_0,\omega}). \tag{5.47}$$

Since $\mathfrak{M}_{\lambda,\rho}$ is invariant under $u_{\lambda,\rho}$ for all $t > 0$ and $h_\lambda = h_{\lambda,\rho}$, we also have for each $\omega \in \Omega$, $t \ge 0$, and $u_0 \in \mathfrak{M}_\lambda^{\mathrm{loc}}(\omega)$ that

$$u_{\lambda,\rho}(t, \omega; u_0) = P_c u_{\lambda,\rho}(t, \omega; u_0) + h_\lambda(P_c u_{\lambda,\rho}(t, \omega; u_0), \theta_t\omega). \tag{5.48}$$

Therefore, for each $u_0 \in \mathring{\mathfrak{M}}_\lambda^{\mathrm{loc}}(\omega)$ and $t \in [0, t^*_{u_0,\omega})$ it holds that

$$u_\lambda(t, \omega; u_0) = P_c u_\lambda(t, \omega; u_0) + h_\lambda(P_c u_\lambda(t, \omega; u_0), \theta_t\omega). \tag{5.49}$$

Using again $\|P_c u_0\|_\alpha < \delta_\rho(\omega)$ for all $u_0 \in \mathring{\mathfrak{M}}_\lambda^{\mathrm{loc}}(\omega)$, it follows from the continuous dependence of the solutions with respect to the initial data and the continuity of the OU process $t \mapsto z_\sigma(\theta_t\omega)$ that there exists $t^{**}_{u_0,\omega} > 0$ such that

$$\|P_c u_\lambda(t, \omega; u_0)\|_\alpha < \delta_\rho(\theta_t\omega), \quad \forall\, t \in [0, t^{**}_{u_0,\omega}). \tag{5.50}$$

This together with (5.49) implies that

$$u_\lambda(t, \omega; u_0) \in \mathring{\mathfrak{M}}_\lambda^{\mathrm{loc}}(\theta_t\omega), \quad \forall\, t \in [0, t_{u_0,\omega}), \tag{5.51}$$

where

$$t_{u_0,\omega} := \min\{t^*_{u_0,\omega}, t^{**}_{u_0,\omega}\}.$$

The local invariance property of $\mathfrak{M}_\lambda^{\mathrm{loc}}$ follows.

Note also that h_λ clearly satisfies the required conditions in Definition 5.1 since $h_{\lambda,\rho}$ does from Step 2. The proof is now complete.

Chapter 6
Local Stochastic Critical Manifolds: Existence and Approximation Formulas

In this chapter, we focus on a typical situation where the control parameter λ varies in an interval that contains the critical value, λ_c, at which the trivial steady state of Eq. (3.1) changes its linear stability stated as a *principle of exchange of stabilities* (*PES*) given in (6.4) below. We show in Lemma 6.1 that this PES implies that the uniform spectrum decomposition (3.11) introduced in Chap. 3 is naturally satisfied. It allows us in turn to establish in Proposition 6.1 the existence and smoothness of a family of local stochastic critical manifolds in the sense of Definition 6.1, which are built—by relying on Chap. 5—as random graphs over some deterministic neighborhoods of the origin in the subspace spanned by the critical modes that lose their stability as λ crosses λ_c. By construction, these manifolds carry nonlinear dynamical information associated with this loss of linear stability.

We then derive in Theorem 6.1 and Corollary 6.1 our main results concerning explicit random approximation formulas to the leading order of these local stochastic critical manifolds about the basic state.

A pullback characterization of the approximation formulas of these stochastic critical manifolds is derived in Volume II [37, Sect. 4.1]. As explained in Volume II, this pullback characterization provides geometric insights of approximating manifolds that is furthermore useful to build up more general manifolds for the small-scale parameterization problem in a noisy environment.

6.1 Standing Hypotheses

Consider the SPDE (3.1). We recall and reformulate some of our assumptions from Chap. 3. The assumptions about the linear operator L_λ are those of Chap. 3. The nonlinearity $F \colon \mathcal{H}_\alpha \to \mathcal{H}$ is assumed here to be C^p-smooth and to take the following form:

$$F(u) = F_k(\underbrace{u, \ldots, u}_{k \text{ times}}) + O(\|u\|_\alpha^{k+1}), \tag{6.1}$$

M.D. Chekroun et al., *Approximation of Stochastic Invariant Manifolds*,
SpringerBriefs in Mathematics, DOI 10.1007/978-3-319-12496-4_6

where $\mathscr{H}_\alpha$ is as before the space associated with the fractional power A^α for some $\alpha \in [0, 1)$, k and p are integers such that

$$p > k \geq 2, \tag{6.2}$$

and

$$F_k \colon \underbrace{\mathscr{H}_\alpha \times \cdots \times \mathscr{H}_\alpha}_{k \text{ times}} \to \mathscr{H} \tag{6.3}$$

is a continuous k-linear operator. Without any confusion we will often write $F_k(u)$ instead of $F_k(u, \ldots, u)$ to simplify the presentation. Note that (6.1) encompasses a large class of nonlinearities satisfying (3.5) and (3.6).

Throughout this chapter, we also assume that the RPDE (4.1) associated with the SPDE (3.1) has a unique classical solution for any given initial datum in $\mathscr{H}_\alpha$ in the sense specified in Proposition 3.1. We denote as before the corresponding RDS acting on $\mathscr{H}_\alpha$ by S_λ. The resulting RDS associated with Eq. (3.1) via the cohomology relation (3.41) is denoted by $\widehat{S}_\lambda$.

We assume furthermore that there exists a critical value λ_c and an integer $m > 0$ such that the following principle of exchange of stabilities (PES) holds at λ_c for the spectrum $\sigma(L_\lambda)$:

$$\begin{aligned} &\operatorname{Re}\beta_j(\lambda) \begin{cases} < 0 & \text{if } \lambda < \lambda_c, \\ = 0 & \text{if } \lambda = \lambda_c, \\ > 0 & \text{if } \lambda > \lambda_c, \end{cases} && \forall\, j \in \{1, \ldots, m\}, \\ &\operatorname{Re}\beta_j(\lambda_c) < 0, && \forall\, j \geq m+1. \end{aligned} \tag{6.4}$$

Note that the critical value λ_c in (6.4) corresponds to the value of λ at which the trivial steady state of Eq. (3.1) (or Eq. (4.1)) changes its linear stability.

In what follows we will make use of the following decomposition of $\sigma(L_\lambda)$ associated with the PES condition (6.4):

$$\begin{aligned} \sigma_{\mathfrak{c}}(L_\lambda) &:= \{\beta_j(\lambda) \mid j = 1, 2, \ldots, m\}, \\ \sigma_{\mathfrak{s}}(L_\lambda) &:= \{\beta_j(\lambda) \mid j = m+1, m+2, \ldots\}, \end{aligned} \tag{6.5}$$

where m is as given in (6.4).

Recall that according to the convention adopted in this monograph, eigenvalues are counted with multiplicity; so repetitions are allowed in (6.5). In particular $\sigma_{\mathfrak{c}}(L_\lambda)$ may, for instance, be constituted of only one eigenvalue of multiplicity m; such situations will be analyzed in details in [36].

6.2 Existence of Local Stochastic Critical Manifolds

Under the assumptions given in the previous section, we establish in Proposition 6.1 below the existence and smoothness of local stochastic (resp. random) invariant manifolds for the RDSs associated with Eq. (3.1) (resp. Eq. (4.1)) for λ in some neighborhood of the critical value λ_c. The resulting manifolds will be called local stochastic (resp. random) critical manifolds; see Definition 6.1.

First, let us introduce the following lemma.

Lemma 6.1 *Assume that the standing hypotheses of Sect.* 6.1 *about the linear operator* L_λ *and its spectrum hold.*

Then, for any $r \in \mathbb{N}^*$, *there exists an open interval* Λ_r *containing the critical value* λ_c *specified in* (6.4), *such that*

$$0 > r\eta_{\mathfrak{c}}(r) > \eta_{\mathfrak{s}}(r), \tag{6.6}$$

where

$$\eta_{\mathfrak{c}}(r) := \inf_{\lambda \in \Lambda_r} \inf_{j=1,\ldots,m} \{\operatorname{Re} \beta_j(\lambda)\}, \quad \eta_{\mathfrak{s}}(r) := \sup_{\lambda \in \Lambda_r} \sup_{j \geq m+1} \{\operatorname{Re} \beta_j(\lambda)\}. \tag{6.7}$$

Remark 6.1

(1) It follows from (6.6) that $\eta_{\mathfrak{s}}(r) < \eta_{\mathfrak{c}}(r)$. This shows in particular that the PES condition implies that L_λ satisfies the uniform spectrum decomposition (3.11) over Λ_r for each $r \in \mathbb{N}^*$. The spectrum of L_λ is thus separated into two disjoint parts, $\sigma_{\mathfrak{c}}(L_\lambda)$ and $\sigma_{\mathfrak{s}}(L_\lambda)$ defined in (6.5), with $\text{card}(\sigma_c(L_\lambda)) = m$, as λ varies in Λ_r. This allows us to introduce subspaces $\mathcal{H}^{\mathfrak{c}}(\lambda)$ and $\mathcal{H}_\alpha^{\mathfrak{s}}(\lambda)$ associated with the decomposition (6.5) for each $\lambda \in \Lambda_r$ as done in (3.18) and (3.20) with therefore $\mathcal{H}^{\mathfrak{c}}(\lambda)$ of fixed dimension m. For the same reasons as given in Chap. 3, we will omit hereafter to point out the λ-dependence of these subspaces.
In the sequel, $\mathcal{H}^{\mathfrak{c}}$ (resp. $\mathcal{H}_\alpha^{\mathfrak{s}}$) will be called the critical subspace (resp. non-critical subspace) of $\mathcal{H}_\alpha$. The corresponding eigenvectors that span the subspace $\mathcal{H}^{\mathfrak{c}}$ (resp. $\mathcal{H}_\alpha^{\mathfrak{s}}$) will be called the critical modes (resp. non-critical modes).

(2) It is important to note also that condition (6.6)—and thus the PES condition in virtue of Lemma 6.1—prevents eigenvalues from $\sigma_{\mathfrak{s}}(L_\lambda)$ given in (6.5) to cross the imaginary axis as λ varies in Λ_r for any $r \in \mathbb{N}^*$. Hence, no eigenvalues other than those of $\sigma_{\mathfrak{c}}(L_\lambda)$ change sign in each such Λ_r. The approximation formulas provided in Theorem 6.1 and Corollary 6.1 are subject to condition (6.6) with $r = 2k$, where k is the leading order of F; see (6.1).

(3) Note that conditions similar to (6.6) are often encountered in the literature when dealing with regularity properties of (deterministic) invariant manifolds; see for instance [45, 84, 107]. Lemma 6.1 shows that such regularity properties may be in fact derived from the PES condition which is easier to check than (6.6) in practice.

Proof It follows from the PES condition (6.4) that

$$\tau := \operatorname{Re} \beta_{m+1}(\lambda_c) < 0. \tag{6.8}$$

Since L_λ is a family of closed operators on $\mathscr{H}$ which depends continuously on λ, it follows that each eigenvalue $\beta_n(\lambda)$ depends continuously on λ; see, e.g., [104, Theorem IV 3.16]. This together with (6.8) implies that there exists an open bounded interval Λ containing λ_c such that

$$\sup_{\lambda \in \Lambda} \operatorname{Re} \beta_{m+1}(\lambda) \leq \frac{3}{4}\tau. \tag{6.9}$$

Then, according to the ordering of the eigenvalues as specified in (3.8)–(3.10), we have

$$\sup_{\lambda \in \Lambda} \sup_{j \geq m+1} \operatorname{Re} \beta_j(\lambda) = \sup_{\lambda \in \Lambda} \operatorname{Re} \beta_{m+1}(\lambda) \leq \frac{3}{4}\tau. \tag{6.10}$$

Since $\operatorname{Re} \beta_j(\lambda_c) = 0$ for $j \in \{1, \ldots, m\}$ (from the PES condition), by the continuous dependence of the corresponding eigenvalues on λ again, we have for each given $r \in \mathbb{N}^*$ that there exists a bounded open interval $\widetilde{\Lambda}_r$ containing λ_c such that

$$\inf_{\lambda \in \widetilde{\Lambda}_r} \inf_{j=1,\ldots,m} \operatorname{Re} \beta_j(\lambda) \geq \frac{\tau}{4r}. \tag{6.11}$$

Now, by introducing

$$\Lambda_r := \Lambda \cap \widetilde{\Lambda}_r,$$

it follows from (6.10) and (6.11) that

$$\begin{aligned} \eta_{\mathfrak{c}}(r) &= \inf_{\lambda \in \Lambda_r} \inf_{j=1,\ldots,m} \{\operatorname{Re} \beta_j(\lambda)\} \geq \frac{\tau}{4r}, \\ \eta_{\mathfrak{s}}(r) &= \sup_{\lambda \in \Lambda_r} \sup_{j \geq m+1} \{\operatorname{Re} \beta_j(\lambda)\} \leq \frac{3\tau}{4}. \end{aligned} \tag{6.12}$$

Recalling that $\tau < 0$ from (6.8), we have that $\eta_{\mathfrak{s}}(r) \leq \frac{3\tau}{4} < \frac{\tau}{4} \leq r\eta_{\mathfrak{c}}(r)$. Note also from the PES condition (6.4) that $\operatorname{Re} \beta_j(\lambda) < 0$ for $j \in \{1, \ldots, m\}$ if $\lambda < \lambda_c$, so that $\eta_{\mathfrak{c}}(r)$ as given in (6.12) is negative. The proof is complete. □

Proposition 6.1 *Assume that the standing hypotheses of Sect.* 6.1 *hold. For each* $r \in \mathbb{N}^*$, *let* Λ_r *be the open interval provided by Lemma* 6.1 *with* $\eta_{\mathfrak{s}}(r)$ *and* $\eta_{\mathfrak{c}}(r)$ *as specified therein which in particular satisfy*

$$0 > r\eta_{\mathfrak{c}}(r) > \eta_{\mathfrak{s}}(r). \tag{6.13}$$

For each $\rho > 0$, let F_ρ be the modified nonlinearity associated with F as given by (5.2).

Let us also introduce:

$$r^* := \min\{r, p\}, \tag{6.14}$$

where p indicates the regularity of F given in Sect. 6.1 *as a C^p-smooth function.*

Then for each

$$\eta \in \Big(\frac{\eta_{\mathfrak{s}}(r)}{r}, \eta_{\mathfrak{c}}(r)\Big), \tag{6.15}$$

there exists $\rho^* > 0$ such that for each $\rho \in (0, \rho^*)$ the following uniform spectral gap conditions hold:

$$\Upsilon_1(F_\rho) = K Lip(F_\rho)\big((\eta_1 - \eta)^{-1} + \Gamma(1-\alpha)(\eta - \eta_2)^{\alpha-1}\big) < \frac{1}{2}, \tag{6.16}$$

and

$$\Upsilon_j(F_\rho) = K Lip(F_\rho)\big((\eta_1 - j\eta)^{-1} + \Gamma(1-\alpha)(j\eta - \eta_2)^{\alpha-1}\big) < 1, \quad \forall j \in \{2, \ldots, r^*\}, \tag{6.17}$$

where K is as given in the partial-dichotomy estimates (3.24).

Moreover, for each $\rho \in (0, \rho^*)$ and each $\lambda \in \Lambda_r$ the following assertions hold:

(i) *The RDS, S_λ, associated with Eq.* (4.1) *admits an m-dimensional local random invariant C^{r^*}–manifold, $\mathfrak{M}_\lambda^{\mathrm{loc}}$, provided by Theorem* 5.1, *which is given by:*

$$\mathfrak{M}_\lambda^{\mathrm{loc}}(\omega) := \{\xi + h_\lambda(\xi, \omega) \mid \xi \in \mathfrak{B}_\rho(\omega)\}, \quad \omega \in \Omega, \tag{6.18}$$

where h_λ and $\mathfrak{B}_\rho$ are as specified therein.

(ii) *The RDS, $\widehat{S}_\lambda$, associated with Eq.* (3.1) *via the cohomology relation* (3.41) *admits an m-dimensional local stochastic invariant C^{r^*}–manifold, $\widehat{\mathfrak{M}}_\lambda^{\mathrm{loc}}$, provided by Corollary* 5.1, *which is given by:*

$$\widehat{\mathfrak{M}}_\lambda^{\mathrm{loc}}(\omega) := \{\xi + \widehat{h}_\lambda(\xi, \omega) \mid \xi \in \widehat{\mathfrak{B}}_\rho\}, \quad \omega \in \Omega. \tag{6.19}$$

Remark 6.2 In case (i), recall from Remark 4.2 (1) and the proof of Theorem 5.1 (Step 3), that the local random invariant manifold, $\mathfrak{M}_\lambda^{\mathrm{loc}}$, is characterized as the random set consisting of all elements u_0 in $\mathcal{H}_\alpha$ such that there exists a complete trajectory of the truncated version of Eq. (4.1) given by Eq. (5.29), passing through u_0 at $t = 0$ and which has a controlled growth as $t \to -\infty$. This control growth is given by $e^{\eta t - \int_t^0 z_\sigma(\theta_\tau \omega)\, d\tau}$ which bounds the $\mathcal{H}_\alpha$-norm of such a trajectory as $t \to -\infty$, for $\eta \in (\frac{\eta_{\mathfrak{s}}(r)}{r}, \eta_{\mathfrak{c}}(r))$ chosen such that the conditions (6.16) and (6.17) hold. Similar statement holds in case (ii).

Proof Lemma 6.1 shows that the standing hypotheses of Sect. 6.1 about the linear operator L_λ imply that the uniform spectrum decomposition (3.11) holds over Λ_r

for each $r \in \mathbb{N}^*$. This proposition is thus a direct consequence of Theorem 5.1 and Corollary 5.1. The C^{r^*}–smoothness of the manifolds is achieved by noting that for each η as given by (6.15), we can choose η_1 and η_2 such that

$$0 > \eta_{\mathfrak{c}}(r) > \eta_1 > \eta_2 > \eta_{\mathfrak{s}}(r), \quad 0 > r\eta_1 > \eta_2, \tag{6.20}$$

and

$$\eta \in \Big(\frac{\eta_2}{r}, \eta_1\Big), \tag{6.21}$$

which implies that $\eta_2 < r\eta < \eta < \eta_1$ (since $\eta < 0$) leading in turn to

$$\eta_2 < j\eta < \eta_1, \quad \forall\, j \in \{1, \ldots, r^*\}, \tag{6.22}$$

so that condition (5.4) of Theorem 5.1 is satisfied. □

This proposition allows us to introduce the following notion of local stochastic (resp. random) critical manifold associated with the SPDE (3.1) (resp. the RPDE (4.1)).

Definition 6.1 For each Λ_r provided by Lemma 6.1 and each $\lambda \in \Lambda_r$, a local random invariant C^{r^*}–manifold associated with Eq. (4.1) guaranteed by Proposition 6.1 is called a local random critical C^{r^*}–manifold, and is denoted by $\mathfrak{M}_\lambda^{\text{crit}}$. The corresponding local invariant manifold function is called a local critical manifold function.

Similarly, for each $\lambda \in \Lambda_r$, a local stochastic invariant C^{r^*}–manifold associated with Eq. (3.1) guaranteed by Proposition 6.1 is called a local stochastic critical C^{r^*}–manifold, and is denoted by $\widehat{\mathfrak{M}}_\lambda^{\text{crit}}$.

Hereafter, a local stochastic (resp. random) critical manifold will be simply referred to as a local critical manifold for the sake of concision.

Remark 6.3 Note that in the deterministic case, the notion of a critical manifold has been used in [92, Sect. 6.3] to refer to a classical center manifold, i.e., only when $\lambda = \lambda_c$. Our notion of critical manifold includes but is not limited to center manifold. For instance, when $\lambda > \lambda_c$ a critical manifold corresponds to an unstable manifold. When $\lambda < \lambda_c$, it corresponds to the manifold constituted by initial data in $\mathcal{H}_\alpha$ from which emanate solutions that are pullback attracted by the origin, whose rate of attraction is entirely determined by the decay rate of $\|P_{\mathfrak{c}} u\|_\alpha$ to zero. These manifolds constitute thus, in a certain sense, the natural objects that reflect locally the PES at the level of the nonlinear dynamics. In [36], it will be shown that the stochastic parameterizing manifolds introduced in Volume II [37] can serve to provide global analogues of critical manifolds that will turn out to be particularly useful in the study of stochastic bifurcations arising in SPDEs driven by (linear) multiplicative noise.

6.3 Approximation of Local Stochastic Critical Manifolds

We present in this section approximation formulas of the local critical manifolds provided by Proposition 6.1.

First, let us introduce the following Landau notations for random functions.

Landau notations. Let X, Y_1, and Y_2 be real Banach spaces, and $\{\Omega, \mathcal{F}, \mathbb{P}\}$ be a probability space. Consider two measurable mappings $f_i : X \times \Omega \to Y_i$, $i = 1, 2$. We write

$$f_1(\xi, \omega) = O(f_2(\xi, \omega)) \tag{6.23}$$

to mean that there exist a constant $M > 0$ and a random open ball $B(0, r(\omega)) \subset X$ centered at 0, such that the following holds

$$\| f_1(\xi, \omega)\|_{Y_1} \leq M \| f_2(\xi, \omega)\|_{Y_2}, \quad \forall\, \xi \in B(0,\ r(\omega)),\ \omega \in \Omega.$$

We denote by

$$f_1(\xi, \omega) = o(f_2(\xi, \omega)) \tag{6.24}$$

to mean that for each constant $\varepsilon > 0$, there exists a random open ball $B(0, r_\varepsilon(\omega)) \subset X$ centered at 0, such that

$$\| f_1(\xi, \omega)\|_{Y_1} \leq \varepsilon \| f_2(\xi, \omega)\|_{Y_2}, \quad \forall\, \xi \in B(0,\ r_\varepsilon(\omega)),\ \omega \in \Omega.$$

Similar notations apply to the case when f_2 is a deterministic mapping and f_2 is still random. Note that when f_1 is also deterministic, these notations agree with the classical Landau notations. It should be clear from the context which one is meant.

We turn now to the main abstract results of this monograph. For this purpose, let us first introduce the following *Lyapunov-Perron integral*:

$$\mathfrak{I}_\lambda(\xi, \omega) = \int_{-\infty}^{0} e^{\sigma(k-1)W_s(\omega)\mathrm{Id}} e^{-sL_\lambda} P_{\mathfrak{s}} F_k(e^{sL_\lambda}\xi)\, \mathrm{d}s, \quad \forall\, \xi \in \mathcal{H}^{\mathfrak{c}},\ \omega \in \Omega. \tag{6.25}$$

Note that this integral is well defined for all $\lambda \in \Lambda$ if $\Lambda \subset \mathbb{R}$ is chosen such that the following condition holds:

$$\eta_{\mathfrak{s}} < \eta_{\mathfrak{c}} \quad \text{and} \quad \eta_{\mathfrak{s}} < k\eta_{\mathfrak{c}}, \tag{6.26}$$

where $\eta_{\mathfrak{c}}$ and $\eta_{\mathfrak{s}}$ are as given in (3.12), and k denotes the order of the nonlinear terms F_k.

Indeed, for such a Λ, we can find η_1 and η_2 such that

$$\eta_{\mathfrak{s}} < \eta_2 < \eta_1 < \eta_{\mathfrak{c}}, \quad \text{and} \quad \eta_2 < k\eta_1. \tag{6.27}$$

Then, by using the dichotomy estimate (3.24b), we get

$$\Big\| \int_{-\infty}^{0} e^{\sigma(k-1)W_s(\omega)\mathrm{Id}} e^{-sL_\lambda} P_{\mathfrak{s}} F_k(e^{sL_\lambda}\xi)\,\mathrm{d}s \Big\|_\alpha \leq K \int_{-\infty}^{0} \frac{e^{\sigma(k-1)W_s(\omega)-\eta_2 s}}{|s|^\alpha} \Big\| F_k(e^{sL_\lambda}\xi) \Big\|\,\mathrm{d}s.$$

Note also that by continuous k-linearity of F_k and the dichotomy estimate (3.24c), there exists $C > 0$ such that

$$\|F_k(e^{sL_\lambda}\xi)\| \leq C\|e^{sL_\lambda}\xi\|_\alpha^k \leq CK^k e^{k\eta_1 s}\|\xi\|_\alpha^k, \quad \forall\, \xi \in \mathcal{H}^{\mathfrak{c}}.$$

We obtain then

$$\Big\| \int_{-\infty}^{0} e^{\sigma(k-1)W_s(\omega)\mathrm{Id}} e^{-sL_\lambda} P_{\mathfrak{s}} F_k(e^{sL_\lambda}\xi)\,\mathrm{d}s \Big\|_\alpha \leq CK^{k+1} \int_{-\infty}^{0} \frac{e^{\sigma(k-1)W_s(\omega)+(k\eta_1-\eta_2)s}}{|s|^\alpha}\mathrm{d}s\,\|\xi\|_\alpha^k,$$

which is finite thanks to the condition $k\eta_1 > \eta_2$ and the fact that $s \mapsto W_s(\omega)$ has sublinear growth as recalled in Lemma 3.1.

The theorem below shows that the Lyapunov-Perron integral, $\mathfrak{I}_\lambda$, provides in fact the leading-order approximation to the stochastic critical manifold function $\widehat{h}_\lambda$ for $\|\xi\|_\alpha$ sufficiently small,[1] and for Λ chosen to be Λ_{2k} such as provided by Lemma 6.1.

Theorem 6.1 *Assume the standing hypotheses of Sect.* 6.1 *hold. Let k be the leading order of the nonlinearity specified in* (6.1), *and Λ_{2k} be the open interval provided by Lemma* 6.1 *with $r = 2k$.*

Let us define, for each $\lambda \in \Lambda_{2k}$, the mapping $h_\lambda^{\mathrm{app}}\colon \mathcal{H}^{\mathfrak{c}} \times \Omega \to \mathcal{H}_\alpha^{\mathfrak{s}}$ as the following Lyapunov-Perron integral:

$$h_\lambda^{\mathrm{app}}(\xi,\omega) := e^{(k-1)z_\sigma(\omega)} \int_{-\infty}^{0} e^{\sigma(k-1)W_s(\omega)\mathrm{Id}} e^{-sL_\lambda} P_{\mathfrak{s}} F_k(e^{sL_\lambda}\xi)\,\mathrm{d}s,$$
$$\forall\, \xi \in \mathcal{H}^{\mathfrak{c}},\ \omega \in \Omega. \tag{6.28}$$

[1] Note however that the random radius $\widehat{r}_\varepsilon(\omega)$ where the approximation (6.31) holds such as deduced from the proof of Theorem 6.1, is not optimal.

Then the local critical manifold function h_λ associated with $\mathfrak{M}_\lambda^{\rm crit}$ provided by Proposition 6.1(i) *(with $r = 2k$) can be approximated to the leading order, k, by $h_\lambda^{\rm app}$ in the following sense:*

- *For any $\varepsilon > 0$, there exists a random open ball $B_\alpha(0, r_\varepsilon(\omega))$ such that*

$$\|h_\lambda(\xi,\omega) - h_\lambda^{\rm app}(\xi,\omega)\|_\alpha \le \varepsilon \|\xi\|_\alpha^k, \; \xi \in B_\alpha(0, r_\varepsilon(\omega)) \cap \mathscr{H}^{\rm c}, \; \lambda \in \Lambda_{2k}, \; \omega \in \Omega. \tag{6.29}$$

Furthermore $r_\varepsilon(\omega) \le \delta_\rho(\omega)$ for all ω, where δ_ρ is the random variable given by

$$\delta_\rho(\omega) := \frac{e^{-z_\sigma(\omega)}\rho}{1+K}, \quad \omega \in \Omega, \tag{6.30}$$

for ρ sufficiently small.[2]

Similarly, for each $\lambda \in \Lambda_{2k}$, let us define $\widehat{h}_\lambda^{\rm app}(\xi,\omega) := e^{z_\sigma(\omega)} h_\lambda^{\rm app}(e^{-z_\sigma(\omega)}\xi, \omega)$, which leads to

$$\boxed{\widehat{h}_\lambda^{\rm app}(\xi,\omega) = \int_{-\infty}^{0} e^{\sigma(k-1)W_s(\omega){\rm Id}} e^{-sL_\lambda} P_{\mathfrak{s}} F_k(e^{sL_\lambda}\xi)\,{\rm d}s, \quad \forall\, \xi \in \mathscr{H}^{\rm c}, \omega \in \Omega.} \tag{AF}$$

Then, the local critical manifold function $\widehat{h}_\lambda$ associated with $\widehat{\mathfrak{M}}_\lambda^{\rm crit}$ provided by Proposition 6.1(ii) *(with $r = 2k$) can be approximated to the leading order, k, by $\widehat{h}_\lambda^{\rm app}$ in the following sense:*

- *For any $\varepsilon > 0$, there exists a random open ball $B_\alpha(0, \widehat{r}_\varepsilon(\omega))$ such that*

$$\|\widehat{h}_\lambda(\xi,\omega) - \widehat{h}_\lambda^{\rm app}(\xi,\omega)\|_\alpha \le \varepsilon \|\xi\|_\alpha^k, \; \xi \in B_\alpha(0, \widehat{r}_\varepsilon(\omega)) \cap \mathscr{H}^{\rm c}, \; \lambda \in \Lambda_{2k}, \; \omega \in \Omega. \tag{6.31}$$

Furthermore, $\widehat{r}_\varepsilon(\omega) \le \widehat{\delta}_\rho$ for all ω, where $\widehat{\delta}_\rho$ is the deterministic constant given by

$$\widehat{\delta}_\rho := \frac{\rho}{(1+K)}, \tag{6.32}$$

for ρ sufficiently small.[3]

Remark 6.4 Note that the above theorem provides therefore conditions under which the local critical manifold function $\widehat{h}_\lambda$ associated with $\widehat{\mathfrak{M}}_\lambda^{\rm crit}$ is approximated by $\widehat{h}_\lambda^{\rm app}$ given by (AF), in the sense that

$$\|\widehat{h}_\lambda(\xi,\omega) - \widehat{h}_\lambda^{\rm app}(\xi,\omega)\|_\alpha = o(\|\xi\|_\alpha^k), \quad \forall\, \lambda \in \Lambda_{2k}, \tag{6.33}$$

where $o(\|\xi\|_\alpha^k)$ is the Landau notation defined in (6.24).

---

[2] More precisely, for $\rho \in (0, \overline{\rho})$ where $\overline{\rho}$ is defined in (6.64) in the proof below.

[3] Same remark given above applies here.

It is important to note that neither these conditions nor the radius over which such an approximation is valid are optimal here. In particular, the deterministic bound provided by (6.32) is artificial and results from the techniques adopted in the proof of Theorem 6.1; see also Remark 5.2.

We assume for the rest of this section that L_λ is self-adjoint. In this case, as we will see the approximation formulas h_λ^{app} and $\widehat{h}_\lambda^{\text{app}}$ given respectively in (6.28) and (AF) can be written as random homogeneous polynomials of order k in the critical state variables, and thus constitute genuine leading-order Taylor approximations of the corresponding families of local critical manifolds.

For that purpose, first note that L_λ being self-adjoint, all its eigenvalues are real. Since $\mathcal{H}_1$, the domain of L_λ, is compactly and densely embedded into $\mathcal{H}$, it follows that the set of normalized eigenvectors $\{e_i \mid i \in \mathbb{N}^*\}$[4] forms a Hilbert basis of $\mathcal{H}$.[5] Then, the critical subspace $\mathcal{H}^{\mathfrak{c}}$ and the non-critical subspace $\mathcal{H}_\alpha^{\mathfrak{s}}$ are given respectively by:

$$\begin{aligned}\mathcal{H}^{\mathfrak{c}} &:= \text{span}\{e_i \mid i = 1, \ldots, m\}, \text{ and}\\ \mathcal{H}_\alpha^{\mathfrak{s}} &:= \overline{\text{span}\{e_j \mid j = m+1, m+2, \ldots, \}}^{\mathcal{H}_\alpha}.\end{aligned} \tag{6.34}$$

Let us also introduce the following notion of random homogeneous polynomials adapted to our framework, which allows us to make precise what we mean by Taylor approximations of critical manifolds.

Definition 6.2 Let $\mathcal{H}^{\mathfrak{c}}$ be the m-dimensional critical subspace and $\mathcal{H}_\alpha^{\mathfrak{s}}$ be the non-critical subspace as given in (6.34). A random homogeneous polynomial function of order p in the critical variable ξ with range in $\mathcal{H}_\alpha^{\mathfrak{s}}$ is a function of the form

$$g\colon \mathcal{H}^{\mathfrak{c}} \times \Omega \to \mathcal{H}_\alpha^{\mathfrak{s}}, \quad (\xi, \omega) \mapsto \sum_{n=m+1}^{\infty} g_n(\xi, \omega) e_n, \tag{6.35}$$

with $g_n(\xi, \cdot) := \langle g(\xi, \cdot), e_n \rangle$,[6] which satisfies furthermore the following conditions:

(i) For each fixed $\xi \in \mathcal{H}^{\mathfrak{c}}$, $g_n(\xi, \cdot)\colon \Omega \to \mathbb{R}$ is $\big(\mathcal{F}; \mathcal{B}(\mathbb{R})\big)$-measurable, where $\mathcal{F}$ is the σ-algebra associated with the Wiener process;
(ii) $g_n(\cdot, \omega)\colon \mathcal{H}^{\mathfrak{c}} \to \mathbb{R}$ is a homogeneous polynomial in $\xi_1, \ldots, \xi_m$ of order p for all ω, where $\xi_i = \langle \xi, e_i \rangle$ for $i = 1, \ldots, m$.

The approximation formula $\widehat{h}_\lambda^{\text{app}}$ (resp. h_λ^{app}) given in (AF) (resp. (6.28)) will be called the leading-order Taylor approximation of the corresponding local critical

[4]As done for the subspaces $\mathcal{H}^{\mathfrak{c}}$ and $\mathcal{H}_\alpha^{\mathfrak{s}}$, we suppress the λ-dependence of the eigenvectors e_i.

[5]This is a direct consequence of the spectral properties of symmetric compact operators (see, e.g., [72, Appendix E, Theorem 7]), and the fact that there exists a positive constant a such that $(-L_\lambda + a\,\text{Id})^{-1}$ exists and is a symmetric compact operator on $\mathcal{H}$.

[6]Where $\langle \cdot, \cdot \rangle$ denoting the inner-product in the ambient Hilbert space $\mathcal{H}$.

manifold function $\widehat{h}_\lambda$ (resp. h_λ) if the estimate (6.31) (resp. (6.29)) holds and if $\widehat{h}_\lambda^{\mathrm{app}}$ (resp. h_λ^{app}) is furthermore a random homogeneous polynomial of order k in the critical state variable $\xi \in \mathcal{H}^{\mathfrak{c}}$.

Corollary 6.1 *Assume the standing hypotheses of Sect. 6.1 hold. Assume furthermore that L_λ is self-adjoint. Then, the approximation h_λ^{app} given in (6.28) constitutes the leading-order Taylor approximation of the corresponding h_λ. More precisely, this approximation is given by:*

$$h_\lambda^{\mathrm{app}}(\xi, \omega) = \sum_{n=m+1}^{\infty} h_\lambda^{\mathrm{app},n}(\xi, \omega) e_n, \quad \forall\, \xi \in \mathcal{H}^{\mathfrak{c}},\ \omega \in \Omega, \tag{6.36}$$

where m is the dimension of the critical subspace $\mathcal{H}^{\mathfrak{c}}$, and $h_\lambda^{\mathrm{app},n}(\xi, \omega)$ is given for each $n \geq m+1$ by

$$h_\lambda^{\mathrm{app},n}(\xi, \omega) := e^{(k-1)z_\sigma(\omega)} \sum_{(i_1,\dots,i_k)\in\mathcal{I}^k} \xi_{i_1}, \dots, \xi_{i_k} \langle F_k(e_{i_1}, \dots, e_{i_k}), e_n\rangle M_n^{i_1,\dots,i_k}(\omega, \lambda). \tag{6.37}$$

Here, $\mathcal{I} = \{1, \dots, m\}$, $\xi_i = \langle \xi, e_i\rangle$ for each $i \in \mathcal{I}$ with $\langle \cdot, \cdot\rangle$ denoting the inner-product in the ambient Hilbert space $\mathcal{H}$, and for each $(i_1, \dots, i_k) \in \mathcal{I}^k$, the $M_n^{i_1,\dots,i_k}(\omega, \lambda)$-term is given by:

$$\boxed{M_n^{i_1,\dots,i_k}(\omega, \lambda) := \int_{-\infty}^{0} e^{\left(\sum_{j=1}^{k} \beta_{i_j}(\lambda) - \beta_n(\lambda)\right)s + \sigma(k-1)W_s(\omega)}\, \mathrm{d}s,} \tag{6.38}$$

where each $\beta_{i_j}(\lambda)$ denotes the eigenvalue associated with the corresponding mode e_{i_j} in $\mathcal{H}^{\mathfrak{c}}$, and $\beta_n(\lambda)$ denotes the eigenvalue associated with the high mode e_n in $\mathcal{H}^{\mathfrak{s}}$.

Under the same assumptions, the approximation $\widehat{h}_\lambda^{\mathrm{app}}$ given in (AF) *constitutes the leading-order Taylor approximation of the corresponding stochastic manifold function, $\widehat{h}_\lambda$. More precisely, this approximation is given by:*

$$\widehat{h}_\lambda^{\mathrm{app}}(\xi, \omega) = \sum_{n=m+1}^{\infty} \widehat{h}_\lambda^{\mathrm{app},n}(\xi, \omega) e_n, \quad \forall\, \xi \in \mathcal{H}^{\mathfrak{c}},\ \omega \in \Omega, \tag{6.39}$$

where $\widehat{h}_\lambda^{\mathrm{app},n}(\xi, \omega)$ is given for each $n \geq m+1$ by

$$\widehat{h}_\lambda^{\mathrm{app},n}(\xi, \omega) := \sum_{(i_1,\dots,i_k)\in\mathcal{I}^k} \xi_{i_1}, \dots, \xi_{i_k} \langle F_k(e_{i_1}, \dots, e_{i_k}), e_n\rangle M_n^{i_1,\dots,i_k}(\omega, \lambda). \tag{6.40}$$

Remark 6.5

(1) Note that $M_n^{i_1,\ldots,i_k}(\omega,\lambda)$ is finite for each $(i_1,\ldots,i_k) \in \mathscr{I}^k$, $\lambda \in \Lambda_{2k}$, and $\omega \in \Omega$. This follows from the fact that $\lim_{t\to\pm\infty} \frac{W_t(\omega)}{t} = 0$ for each ω (see Lemma 3.1 again), and that $\sum_{j=1}^k \beta_{i_j}(\lambda) > \beta_n(\lambda)$. This latter inequality holds since according to (6.6) and (6.7), we have that $\sum_{j=1}^k \beta_{i_j}(\lambda) \geq k\eta_{\mathfrak{c}}(2k) > 2k\eta_{\mathfrak{c}}(2k) > \eta_{\mathfrak{s}}(2k) \geq \beta_n(\lambda)$ for all $\lambda \in \Lambda_{2k}$. Besides, it can be checked that $M_n^{i_1,\ldots,i_k}(\omega,\lambda)$ is a tempered random variable. In Volume II [37, Chap. 5], it will be shown that such random variables convey memory effects that will be expressed in terms of decay of correlations provided that σ lives in some admissible range; see [37, Lemma 5.1].

(2) It is worth noting that the $M_n^{i_1,\ldots,i_k}(\omega,\lambda)$-terms come with $\langle F_k(e_{i_1},\ldots,e_{i_k}), e_n\rangle$, i.e., the nonlinear, leading-order self-interactions of the (corresponding) critical modes, as projected against the non-critical mode e_n. Note that by definition, any permutation of a k-tuple $(i_1,\ldots,i_k)$ corresponds to a same $M_n^{i_1,\ldots,i_k}(\omega,\lambda)$-term but may correspond to different interactions projected against the e_n-mode.

(3) Note that if $\sigma_{\mathfrak{c}}(L_\lambda)$ as defined in (6.5) consists of one eigenvalue with multiplicity m, i.e.,

$$\beta_1(\lambda) = \cdots = \beta_m(\lambda) =: \beta_*(\lambda), \ \forall\, \lambda \in \Lambda_{2k}, \tag{6.41}$$

then each $M_n^{i_1,\ldots,i_k}(\omega,\lambda)$-term reduces to the following simplified form:

$$M_n(\omega,\lambda) := \int_{-\infty}^{0} e^{[k\beta_*(\lambda)-\beta_n(\lambda)]s+\sigma(k-1)W_s(\omega)}\, ds. \tag{6.42}$$

(4) When $\sigma = 0$, in the general case for L_λ, the approximation result stated in Theorem 6.1 recovers those obtained in [92, Lemma 6.2.4], [117, Chap. 3], and [120, Appendix A]. The proof presented in Sect. 6.4 below works literally for this case by simply setting σ and hence z_σ to zero; see also [37, Lemma 4.1]. In the self-adjoint case for L_λ, the approximation formula (6.39) becomes deterministic when $\sigma = 0$, where the M_n-terms given in (6.38) are then reduced to

$$M_n^{i_1,\ldots,i_k}(\lambda) = \int_{-\infty}^{0} e^{\left(\sum_{j=1}^k \beta_{i_j}(\lambda)-\beta_n(\lambda)\right)s}\, ds = \frac{1}{\sum_{j=1}^k \beta_{i_j}(\lambda) - \beta_n(\lambda)}, \tag{6.43}$$

for each $n \geq m+1$ and $\lambda \in \Lambda_{2k}$. In this case also, the $M_n^{i_1,\ldots,i_k}$-terms come with the nonlinear, leading-order self-interactions of the critical modes, as projected against the non-critical mode e_n.

6.4 Proofs of Theorem 6.1 and Corollary 6.1

We turn first to the description of the main ideas of the proof of Theorem 6.1.

Skeleton of the Proof of Theorem 6.1. We focus on the approximation of h_λ, the results for $\widehat{h}_\lambda$ following then from the cohomology relation given in (5.13).

The estimate given in (6.29) will be achieved by using an appropriate "pivot" quantity $h^{\mathrm{app}}_{\lambda,\rho}$ in the basic triangle inequality:

$$\begin{aligned} &\|h_\lambda(\xi,\omega) - h^{\mathrm{app}}_\lambda(\xi,\omega)\|_\alpha \\ &\quad \le \|h_\lambda(\xi,\omega) - h^{\mathrm{app}}_{\lambda,\rho}(\xi,\omega)\|_\alpha + \|h^{\mathrm{app}}_{\lambda,\rho}(\xi,\omega) - h^{\mathrm{app}}_\lambda(\xi,\omega)\|_\alpha. \end{aligned} \tag{6.44}$$

To describe this pivot quantity we first recall from Chap. 5 that $h_\lambda = h_{\lambda,\rho}$, where $h_{\lambda,\rho}$ is obtained as the fixed point of the following integral equation

$$h_{\lambda,\rho}(\xi,\omega) = \int_{-\infty}^{0} \mathfrak{T}_{\lambda,\sigma}(0,s;\omega) P_{\mathfrak{s}} G_\rho\big(\theta_s\omega, u_{\lambda,\rho}(s,\omega;\xi + h_{\lambda,\rho}(\xi,\omega))\big)\mathrm{d}s \tag{6.45}$$

associated with a modified equation of Eq. (4.1) given by

$$\frac{\mathrm{d}u}{\mathrm{d}t} = L_\lambda u + z_\sigma(\theta_t\omega)u + G_\rho(\theta_t\omega,\, u), \tag{6.46}$$

where $\mathfrak{T}_{\lambda,\sigma}$ is the solution operator introduced in Sect. 3.4, G_ρ is an appropriate cut-off version of G, and $u_{\lambda,\rho}$ is a mild solution to (6.46) on $(-\infty, 0]$ taking value $\xi + h_{\lambda,\rho}(\xi,\omega)$ at $t = 0$; see also (5.29), (5.30) and (4.8).

The pivot $h^{\mathrm{app}}_{\lambda,\rho}$ is then obtained by replacing G_ρ in the integral equation (6.45) with $G_{\rho,k}$ (the leading-order term of G_ρ as defined in (6.56)) and by replacing the mild solution $u_{\lambda,\rho}$ of Eq. (6.46) involved in (6.45) with the backward solution $v(s) := \mathfrak{T}_{\lambda,\sigma}(s,0;\omega)\xi$, $s \le 0$, associated with the random linear equation:

$$\frac{\mathrm{d}v}{\mathrm{d}t} = L_\lambda v + z_\sigma(\theta_t\omega)v. \tag{6.47}$$

Note that $v(s)$ is well-defined since $\xi \in \mathcal{H}^{\mathrm{c}}$; see Sect. 3.4.

In other words, the pivot quantity $h^{\mathrm{app}}_{\lambda,\rho}$ reads:

$$h^{\mathrm{app}}_{\lambda,\rho}(\xi,\omega) := \int_{-\infty}^{0} \mathfrak{T}_{\lambda,\sigma}(0,s;\omega) P_{\mathfrak{s}} G_{\rho,k}(\theta_s\omega, \mathfrak{T}_{\lambda,\sigma}(s,0;\omega)\xi)\mathrm{d}s, \quad \xi \in \mathcal{H}^{\mathrm{c}},\ \omega \in \Omega. \tag{6.48}$$

Let us first describe how we control $\|h^{\mathrm{app}}_{\lambda,\rho}(\xi,\omega) - h^{\mathrm{app}}_{\lambda}(\xi,\omega)\|_\alpha$. Actually, $h^{\mathrm{app}}_{\lambda,\rho}$ differs from the approximation formula $h^{\mathrm{app}}_{\lambda}$ as rewritten in (6.95) by replacing G_k with $G_{\rho,k}$, so that $h^{\mathrm{app}}_{\lambda,\rho} - h^{\mathrm{app}}_{\lambda}$ is given by:

$$\begin{aligned} & h^{\mathrm{app}}_{\lambda}(\xi,\omega) - h^{\mathrm{app}}_{\lambda,\rho}(\xi,\omega) \\ &= \int_{-\infty}^{0} \mathfrak{T}_{\lambda,\sigma}(0,s;\omega)\Big(1-\zeta\Big(\frac{e^{z_\sigma(\theta_s\omega)}\|v(s)\|_\alpha}{\rho}\Big)\Big) P_{\mathfrak{s}} G_k(\theta_s\omega, v(s))\,\mathrm{d}s; \end{aligned}$$

see (6.96).

By noting that $\zeta\left(\frac{e^{z_\sigma(\theta_s\omega)}\|v(s)\|_\alpha}{\rho}\right) = 1$ if $e^{z_\sigma(\theta_s\omega)}\|v(s)\|_\alpha \le \rho$, the last identity above becomes:

$$\begin{aligned} & h^{\mathrm{app}}_{\lambda}(\xi,\omega) - h^{\mathrm{app}}_{\lambda,\rho}(\xi,\omega) \\ &= \int_{-\infty}^{s_0(\xi)} \mathfrak{T}_{\lambda,\sigma}(0,s;\omega)\Big(1-\zeta\Big(\frac{e^{z_\sigma(\theta_s\omega)}\|v(s)\|_\alpha}{\rho}\Big)\Big) P_{\mathfrak{s}} G_k(\theta_s\omega, v(s))\,\mathrm{d}s, \end{aligned}$$

where $s_0(\xi) < 0$ denotes the first time at which the backward solution $v(s)$ of Eq. (6.47) emanating from some $\xi \in \mathcal{H}^{\mathfrak{c}}$ leaves $B_\alpha(0, e^{-z_\sigma(\theta_s\omega)}\rho)$. This remark allows us in Step 5 to control $\|h^{\mathrm{app}}_{\lambda}(\xi,\omega) - h^{\mathrm{app}}_{\lambda,\rho}(\xi,\omega)\|_\alpha$ as follows by usage of the partial-dichotomy estimates (3.46) and continuous k-linear properties of G_k:

$$\|h^{\mathrm{app}}_{\lambda}(\xi,\omega) - h^{\mathrm{app}}_{\lambda,\rho}(\xi,\omega)\|_\alpha \le \int_{-\infty}^{s_0(\xi)} \kappa_0(s,\omega)\mathrm{d}s\,\|\xi\|^k,$$

where $\kappa_0(s,\omega)$ is some exponentially decaying factor; see (6.102).[7]

By taking ξ in a sufficiently small random ball $B_\alpha(0, r_\varepsilon(\omega)) \cap \mathcal{H}^{\mathfrak{c}}$, it is then shown that

$$\int_{-\infty}^{s_0(\xi)} \kappa_0(s,\omega)\mathrm{d}s \le \frac{\varepsilon}{2},$$

leading to a good control of $\|h^{\mathrm{app}}_{\lambda}(\xi,\omega) - h^{\mathrm{app}}_{\lambda,\rho}(\xi,\omega)\|_\alpha$. The last inequality above follows from the observation that $s_0(\xi)$ gets more negative as $\|\xi\|_\alpha$ gets smaller.

The control of $\|h_\lambda(\xi,\omega) - h^{\mathrm{app}}_{\lambda,\rho}(\xi,\omega)\|_\alpha$ is more challenging and is carried out in Steps 2, 3, and 4. Note that according to (6.45), (6.48), and the fact $h_\lambda = h_{\lambda,\rho}$, it holds that

[7] The OU process $z_\sigma(\theta_s\omega)$ is involved in the factor $\kappa_0(s,\omega)$, and its control is made possible thanks to the growth control estimates given in (3.30).

$$h_\lambda(\xi,\omega) - h^{\text{app}}_{\lambda,\rho}(\xi,\omega) = \int_{-\infty}^{0} \mathfrak{T}_{\lambda,\sigma}(0,s;\omega)P_{\mathfrak{s}}\Big(G_\rho(\theta_s\omega,u(s)) - G_{\rho,k}(\theta_s\omega,v(s))\Big)\mathrm{d}s, \tag{6.49}$$

where we have introduced $u(s) := u_{\lambda,\rho}(s,\omega;\xi + h_{\lambda,\rho}(\xi,\omega)$ to simplify the presentation. The main task is then to obtain a good control of the quantity $\|G_\rho(\theta_s\omega,u(s)) - G_{\rho,k}(\theta_s\omega,v(s))\|$, which can be split further as follows by introducing another pivot $G_{\rho,k}(\theta_s\omega,u(s))$:

$$\begin{aligned}\|G_\rho(\theta_s\omega,u(s)) - G_{\rho,k}(\theta_s\omega,v(s))\| \le \|G_\rho(\theta_s\omega,u(s)) - G_{\rho,k}(\theta_s\omega,u(s))\|\\ + \|G_{\rho,k}(\theta_s\omega,u(s)) - G_{\rho,k}(\theta_s\omega,v(s))\|.\end{aligned} \tag{6.50}$$

We then point out in Step 2 at the level of "vector fields" key estimates of $\|G_\rho(\omega,u) - G_{\rho,k}(\omega,u)\|$ and $\|G_{\rho,k}(\omega,u_1) - G_{\rho,k}(\omega,u_2)\|$ that result from continuous k-linear properties of the leading-order term F_k (see again (5.30) and (6.56) for the definitions of the notations):

$$\begin{aligned}\|G_\rho(\omega,u) - G_{\rho,k}(\omega,u)\| &\le C_1(\omega)\|u\|_\alpha^{k+1}, \qquad \forall\, u\in\mathcal{H}_\alpha,\ \omega\in\Omega,\\ \|G_{\rho,k}(\omega,u_1) - G_{\rho,k}(\omega,u_2)\| &\le C_2(\omega)\big(\|u_1\|_\alpha^{k-1} + \|u_2\|_\alpha^{k-1}\big)\|u_1-u_2\|_\alpha\\ &\quad + C_3(\omega)\|u_2\|_\alpha^k\|u_1-u_2\|_\alpha,\ \forall\, u_1,u_2\in\mathcal{H}_\alpha,\ \omega\in\Omega,\end{aligned}$$

where the $C_i(\omega)$'s are some positive random constants; see (6.65) and (6.66). Using these estimates and the partial-dichotomy estimates (3.46) in (6.49), we then obtain within the same step that

$$\|h_\lambda(\xi,\omega) - h^{\text{app}}_{\lambda,\rho}(\xi,\omega)\|_\alpha \le I_1(\xi,\omega) + I_2(\xi,\omega) + I_3(\xi,\omega),$$

where

$$\begin{aligned}I_1(\xi,\omega) &= \int_{-\infty}^{0}\kappa_1(s,\omega)\|u(s)\|_\alpha^{k+1}\mathrm{d}s,\\ I_2(\xi,\omega) &= \int_{-\infty}^{0}\kappa_2(s,\omega)(\|u(s)\|_\alpha^{k-1} + \|v(s)\|_\alpha^{k-1})\|u(s)-v(s)\|_\alpha\mathrm{d}s,\\ I_3(\xi,\omega) &= \int_{-\infty}^{0}\kappa_3(s,\omega)\|v(s)\|_\alpha^{k}\|u(s)-v(s)\|_\alpha\mathrm{d}s,\end{aligned} \tag{6.51}$$

with the $\kappa_i(s,\omega)$'s denoting some exponentially decaying factors, and $u(s)$, $v(s)$ the solutions associated with respectively Eqs. (6.46) and (6.47) as before; see (6.70) for more details. Here, $I_1(\xi,\omega)$ provides a control of

$$\Big\| \int_{-\infty}^{0} \mathfrak{T}_{\lambda,\sigma}(0,s;\omega) P_{\mathfrak{s}} \Big(G_\rho(\theta_s\omega, u(s)) - G_{\rho,k}(\theta_s\omega, u(s)) \Big) \mathrm{d}s \Big\|_\alpha,$$

and $I_1(\xi,\omega) + I_3(\xi,\omega)$ provides a control of

$$\Big\| \int_{-\infty}^{0} \mathfrak{T}_{\lambda,\sigma}(0,s;\omega) P_{\mathfrak{s}} \Big(G_{\rho,k}(\theta_s\omega, u(s)) - G_{\rho,k}(\theta_s\omega, v(s)) \Big) \mathrm{d}s \Big\|_\alpha.$$

In Step 3, the following estimates of $\|u(s)\|_\alpha$, $\|v(s)\|_\alpha$, and $\|u(s) - v(s)\|_\alpha$ are then carried out:

$$\begin{aligned} \|u(s)\|_\alpha &\le \kappa_4(s,\omega)\|\xi\|_\alpha, \\ \|v(s)\|_\alpha &\le \kappa_5(s,\omega)\|\xi\|_\alpha, \\ \|u(s) - v(s)\|_\alpha &\le \kappa_6(s,\omega)\|\xi\|_\alpha^k, \end{aligned} \tag{6.52}$$

where the $\kappa_i(s,\omega)$'s are positive but not necessarily decaying factors; see (6.71)–(6.73).

The estimate for $\|v(s)\|_\alpha$ follows directly from the partial-dichotomy estimate (3.46c). The controls of $\|u(s)\|_\alpha$ and $\|u(s) - v(s)\|_\alpha$ are subject to a control of $\|u(\cdot)\|_{C_\eta^-}$ by $2K\|\xi\|_\alpha$ as pointed out at the beginning of Step 3 resulting from application of the estimate (B.16) derived in the proof of Theorem 4.1. The rest of Step 3 is devoted to showing that

$$\|u(s) - v(s)\|_\alpha \le C_4(\omega)\kappa_7(s,\omega)\|u(\cdot)\|_{C_\eta^-}^k,$$

where $\kappa_7(s,\omega)$ is a positive factor and $C_4(\omega)$ is a positive random constant. This quantity $C_4(\omega)$ is obtained by appropriately controlling the integrands arising from application of the dichotomy estimates to the integral equation (6.54) given below that $u(t)$ satisfies; see (6.76), (6.77) and (6.84). A basic algebraic lemma (Lemma 6.2) is then used to achieve this appropriate control.

In Step 4, by using (6.52) in (6.51), we obtain with the help of Lemma 6.2 that

$$\begin{aligned} &\|h_\lambda(\xi,\omega) - h_{\lambda,\rho}^{\mathrm{app}}(\xi,\omega)\|_\alpha \\ &\quad \le I_1(\xi,\omega) + I_2(\xi,\omega) + I_3(\xi,\omega) \\ &\quad \le C_5(\omega)\|\xi\|_\alpha^{k+1} + C_6(\omega)\|\xi\|_\alpha^{2k-1} + C_7(\omega)\|\xi\|_\alpha^{2k}, \quad \forall\, \xi \in \mathscr{H}^{\mathfrak{c}}, \end{aligned}$$

where the $C_i(\omega)$'s are some positive random constants; see (6.90). It then follows that $\|h_\lambda(\xi,\omega) - h^{\text{app}}_{\lambda,\rho}(\xi,\omega)\|_\alpha \leq \frac{\varepsilon}{2}\|\xi\|^k_\alpha$ when ξ is in some sufficiently small random ball $B_\alpha(0, r_\varepsilon(\omega)) \cap \mathscr{H}^{\mathfrak{c}}$.

Finally, the estimate (6.29) follows by combining estimates for $\|h_\lambda(\xi,\omega) - h^{\text{app}}_{\lambda,\rho}(\xi,\omega)\|_\alpha$ and $\|h^{\text{app}}_{\lambda,\rho}(\xi,\omega) - h^{\text{app}}_{\lambda}(\xi,\omega)\|_\alpha$, which is done in Step 6.

Proof of Theorem 6.1 We proceed in six steps following the ideas outlined above.

Step 1. Integral equation for h_λ and the pivot $h^{\text{app}}_{\lambda,\rho}$. In this step, we recall an integral equation that h_λ satisfies, and introduce an intermediate approximation formula $h^{\text{app}}_{\lambda,\rho}$. Corresponding error estimates between h_λ and $h^{\text{app}}_{\lambda,\rho}$ will be obtained in the next three steps.

First, let us recall that from the proof of Theorem 5.1, we have

$$h_\lambda = h_{\lambda,\rho},$$

where $h_{\lambda,\rho}$ is the global random invariant manifold function associated with the modified equation (5.29). By application of Theorem 4.1 to Eq. (5.29), we know that $h_{\lambda,\rho}$ satisfies the following integral equation:

$$h_{\lambda,\rho}(\xi,\omega) = \int_{-\infty}^{0} \mathfrak{T}_{\lambda,\sigma}(0,s;\omega) P_{\mathfrak{s}} G_\rho\big(\theta_s\omega, u_{\lambda,\rho}(s,\omega;\xi + h_{\lambda,\rho}(\xi,\omega))\big)\,\mathrm{d}s, \quad (6.53)$$

where $\mathfrak{T}_{\lambda,\sigma}$ is as given in Sect. 3.4 and $u_{\lambda,\rho}(\cdot,\omega;\xi + h_{\lambda,\rho}(\xi,\omega))$ is a mild solution of Eq. (5.29) defined on $(-\infty, 0]$ taking value $\xi + h_{\lambda,\rho}(\xi,\omega)$ at $t = 0$. Recall from the construction of such an $h_{\lambda,\rho}$ (see Step 1 in the proof of Theorem 4.1) that the corresponding mild solution $u_{\lambda,\rho}(\cdot,\omega;\xi + h_{\lambda,\rho}(\xi,\omega))$ is obtained as the unique fixed point in C^-_η of the integral operator $\mathscr{N}^{\omega,\lambda}_{\xi,\rho}$ defined by:

$$\begin{aligned}\mathscr{N}^{\omega,\lambda}_{\xi,\rho}[u](s) &:= \mathfrak{T}_{\lambda,\sigma}(s,0;\omega)\xi - \int_{s}^{0} \mathfrak{T}_{\lambda,\sigma}(s,s';\omega) P_{\mathfrak{c}} G_\rho(\theta_{s'}\omega, u(s'))\,\mathrm{d}s' \\ &\quad + \int_{-\infty}^{s} \mathfrak{T}_{\lambda,\sigma}(s,s';\omega) P_{\mathfrak{s}} G_\rho(\theta_{s'}\omega, u(s'))\,\mathrm{d}s', \qquad s \leq 0.\end{aligned} \quad (6.54)$$

This fact will be used in Step 3 below.

Now, for each $\rho > 0$ and $\lambda \in \Lambda_{2k}$, let us introduce a mapping $h^{\text{app}}_{\lambda,\rho}: \mathscr{H}^{\mathfrak{c}} \times \Omega \to \mathscr{H}^{\mathfrak{s}}_\alpha$ defined by:

$$h^{\text{app}}_{\lambda,\rho}(\xi,\omega) := \int_{-\infty}^{0} \mathfrak{T}_{\lambda,\sigma}(0,s;\omega) P_{\mathfrak{s}} G_{\rho,k}(\theta_s\omega, \mathfrak{T}_{\lambda,\sigma}(s,0;\omega)\xi)\,\mathrm{d}s, \quad \xi \in \mathscr{H}^{\mathfrak{c}},\ \omega \in \Omega, \quad (6.55)$$

where

$$G_{\rho,k}(\omega,\, u) := e^{-z_\sigma(\omega)} F_{\rho,k}(e^{z_\sigma(\omega)} u) \quad \text{with} \quad F_{\rho,k}(u) := \zeta\Big(\frac{\|u\|_\alpha}{\rho}\Big) F_k(u), \tag{6.56}$$

and ζ is the cut-off function defined in (5.1).

Compared with $h_{\lambda,\rho}(\xi, \omega)$ defined in (6.53), the nonlinearity G_ρ is replaced here by $G_{\rho,k}$, and the pathwise solution $u_{\lambda,\rho}$ is replaced by the solution $\mathfrak{T}_{\lambda,\sigma}(s, 0; \omega)\xi$ of the linear problem $\frac{\mathrm{d}v}{\mathrm{d}t} = L_\lambda v + z_\sigma(\theta_t\omega)v$ with initial datum $\xi \in \mathcal{H}^{\mathrm{c}}$.

We aim to show in the next three steps that there exists $\overline{\rho} > 0$, such that for each fixed $\rho \in (0, \overline{\rho})$ and any $\varepsilon > 0$, there exists a random open ball $B_\alpha(0, r_\varepsilon^*(\omega)) \subset \mathcal{H}_\alpha$ verifying

$$\begin{aligned} \|h_{\lambda,\rho}(\xi,\omega) - h^{\mathrm{app}}_{\lambda,\rho}(\xi,\omega)\|_\alpha \le \frac{\varepsilon}{2}\|\xi\|_\alpha^k,\ \forall \xi \in B_\alpha(0, r_\varepsilon^*(\omega)) \cap \mathcal{H}^{\mathrm{c}}, \\ \lambda \in \Lambda_{2k},\ \omega \in \Omega. \end{aligned} \tag{6.57}$$

Since in what follows all estimates are uniform in λ for all $\lambda \in \Lambda_{2k}$, and are produced for each fixed ω, to simplify the presentation, we introduce for all $s \le 0$ and $\xi \in \mathcal{H}^{\mathrm{c}}$ the following two notations keeping only the time dependence explicit[8]:

$$\begin{aligned} u(s) &:= u_{\lambda,\rho}(s, \omega; \xi + h_{\lambda,\rho}(\xi,\omega)), \\ v(s) &:= \mathfrak{T}_{\lambda,\sigma}(s, 0; \omega)\xi. \end{aligned} \tag{6.58}$$

Then from (6.53) and (6.55) we obtain

$$\begin{aligned} &h_{\lambda,\rho}(\xi,\omega) - h^{\mathrm{app}}_{\lambda,\rho}(\xi,\omega) \\ &\quad = \int_{-\infty}^{0} \mathfrak{T}_{\lambda,\sigma}(0, s; \omega)\Big(P_{\mathfrak{s}} G_\rho(\theta_s\omega, u(s)) - P_{\mathfrak{s}} G_{\rho,k}(\theta_s\omega, v(s))\Big)\mathrm{d}s. \end{aligned} \tag{6.59}$$

The integral representation of $h_{\lambda,\rho} - h^{\mathrm{app}}_{\lambda,\rho}$ involved in (6.59) motivates the estimates presented in the next three steps, which are organized as follows. In Step 2, we first point out some key estimates related to G_ρ and $G_{\rho,k}$ at the level of "vector fields," which are then used to control (6.59) via three integral terms $I_1(\xi,\omega)$, $I_2(\xi,\omega)$, and $I_3(\xi,\omega)$. These integral terms involve the mild solution $u(s)$ and the linear flow $v(s)$; see (6.70). In Step 3, we derive some related estimates of $\|u(s)\|_\alpha$, $\|v(s)\|_\alpha$, and $\|u(s) - v(s)\|_\alpha$, which will be used in Step 4 to derive the desired error estimate announced in (6.57).

[8] It is also safe to suppress the ρ-dependence of the mild solution $u_{\lambda,\rho}$ because all the estimates on $u_{\lambda,\rho}$ are done for a fixed $\rho \in (0, \overline{\rho})$, with $\overline{\rho}$ specified later in (6.64). The dependence on σ is also removed for the sake of readability.

Step 2. Estimates of the nonlinear terms at the level of "vector fields." In this step, we first establish some estimates about G_ρ and $G_{\rho,k}$ as "vector fields" from $(\mathscr{H}_\alpha, \|\cdot\|_\alpha)$ to $(\mathscr{H}, \|\cdot\|)$, and then derive a control of $\|h_{\lambda,\rho}(\xi,\omega) - h^{\mathrm{app}}_{\lambda,\rho}(\xi,\omega)\|_\alpha$ as given in (6.70).

We first note that

$$\begin{aligned}
\|P_{\mathfrak{s}} G_\rho(\theta_s\omega, u(s)) - P_{\mathfrak{s}} G_{\rho,k}(\theta_s\omega, v(s))\| & \\
\le \|P_{\mathfrak{s}} G_\rho(\theta_s\omega, u(s)) - P_{\mathfrak{s}} G_{\rho,k}(\theta_s\omega, u(s))\| & \qquad (6.60) \\
+ \|P_{\mathfrak{s}} G_{\rho,k}(\theta_s\omega, u(s)) - P_{\mathfrak{s}} G_{\rho,k}(\theta_s\omega, v(s))\|. &
\end{aligned}$$

To proceed further from (6.60), we establish some estimates about F_ρ and $F_{\rho,k}$, where

$$F_\rho(u) := \zeta\Big(\frac{\|u\|_\alpha}{\rho}\Big) F(u), \quad F_{\rho,k}(u) := \zeta\Big(\frac{\|u\|_\alpha}{\rho}\Big) F_k(u), \qquad (6.61)$$

and ζ is still the cut-off function defined in (5.1).

In the sequel, $C > 0$ will denote a generic positive constant which may or may not depend on ρ. We claim that there exist $\widetilde{\rho} > 0$ and a positive constant C such that:

$$\|F_\rho(u)\| \le C\|u\|_\alpha^k, \qquad \forall\, \rho \in (0, \widetilde{\rho}),\ u \in \mathscr{H}_\alpha, \qquad (6.62a)$$

$$\|F_\rho(u) - F_{\rho,k}(u)\| \le C\|u\|_\alpha^{k+1}, \qquad \forall\, \rho \in (0, \widetilde{\rho}),\ u \in \mathscr{H}_\alpha. \qquad (6.62b)$$

Let us first check (6.62a). It follows from the assumption (6.1) that $F(u) = O(\|u\|_\alpha^k)$. Hence, there exists $\widetilde{\rho} > 0$ and a positive constant C such that

$$\|F(u)\| \le C\|u\|_\alpha^k, \qquad \forall\, u \in B_\alpha(0, 2\widetilde{\rho}).$$

Then, for each $\rho \in (0, \widetilde{\rho})$, (6.62a) holds for all $u \in B_\alpha(0, 2\rho)$. Note also that (6.62a) holds obviously if $\|u\|_\alpha \ge 2\rho$ by the definition of the cut-off function ζ in (5.1).

From (6.1), we also have that $F(u) - F_k(u) = O(\|u\|_\alpha^{k+1})$. Following the same arguments as above, we obtain (6.62b) by choosing possibly a smaller $\widetilde{\rho}$ and a larger constant C.

Moreover, since F_k is a continuous k-linear operator and ζ is Lipschitz, we have that for each $\rho > 0$, there exists $C > 0$ such that

$$\begin{aligned}
\|F_{\rho,k}(u_1) - F_{\rho,k}(u_2)\| &= \Big\| \zeta\Big(\frac{\|u_1\|_\alpha}{\rho}\Big) F_k(\underbrace{u_1, \ldots, u_1}_{k \text{ times}}) \\
&\qquad - \zeta\Big(\frac{\|u_2\|_\alpha}{\rho}\Big) F_k(\underbrace{u_2, \ldots, u_2}_{k \text{ times}}) \Big\| \\
&\le \Big\| \zeta\Big(\frac{\|u_1\|_\alpha}{\rho}\Big) F_k(u_1, \ldots, u_1)
\end{aligned}$$

$$
\begin{aligned}
& - \zeta\Big(\frac{\|u_1\|_\alpha}{\rho}\Big) F_k(\underbrace{u_1, \ldots, u_1}_{k-1 \text{ times}}, u_2)\Big\| \\
& + \Big\|\zeta\Big(\frac{\|u_1\|_\alpha}{\rho}\Big) F_k(\underbrace{u_1, \ldots, u_1}_{k-1 \text{ times}}, u_2) \\
& - \zeta\Big(\frac{\|u_1\|_\alpha}{\rho}\Big) F_k(\underbrace{u_1, \ldots, u_1}_{k-2 \text{ times}}, u_2, u_2)\Big\| \\
& + \cdots \\
& + \Big\|\zeta\Big(\frac{\|u_1\|_\alpha}{\rho}\Big) F_k(u_1, u_2, \ldots, u_2) \\
& - \zeta\Big(\frac{\|u_1\|_\alpha}{\rho}\Big) F_k(u_2, \ldots, u_2)\Big\| \\
& + \Big\|\zeta\Big(\frac{\|u_1\|_\alpha}{\rho}\Big) F_k(u_2, \ldots, u_2) \\
& - \zeta\Big(\frac{\|u_2\|_\alpha}{\rho}\Big) F_k(u_2, \ldots, u_2)\Big\| \\
\le\, & C \sum_{i=0}^{k-1} \|u_1\|_\alpha^{k-1-i} \|u_2\|_\alpha^i \|u_1 - u_2\|_\alpha \\
& + C\|u_2\|_\alpha^k \|u_1 - u_2\|_\alpha \\
\le\, & C \sum_{i=0}^{k-1} \big(\|u_1\|_\alpha^{k-1} + \|u_2\|_\alpha^{k-1}\big) \|u_1 - u_2\|_\alpha \\
& + C\|u_2\|_\alpha^k \|u_1 - u_2\|_\alpha \\
=\, & kC\big(\|u_1\|_\alpha^{k-1} + \|u_2\|_\alpha^{k-1}\big) \|u_1 - u_2\|_\alpha \\
& + C\|u_2\|_\alpha^k \|u_1 - u_2\|_\alpha, \ \forall\, u_1, u_2 \in \mathscr{H}_\alpha. \qquad (6.63)
\end{aligned}
$$

Now, let us define $\overline{\rho}$ by:

$$\overline{\rho} := \min\{\widetilde{\rho}, \rho^*\}, \qquad (6.64)$$

where $\widetilde{\rho}$ is as specified in (6.62a) and (6.62b), and ρ^* is as given in Proposition 6.1 to ensure the existence of a critical manifold for $\lambda \in \Lambda_{2k}$; see also Remark 5.2. From now on, we fix an arbitrary $\rho \in (0, \overline{\rho})$.

Recalling from (5.30) and (6.56) that

$$G_\rho(\omega, u) = e^{-z_\sigma(\omega)} F_\rho(e^{z_\sigma(\omega)} u), \quad G_{\rho,k}(\omega, u) = e^{-z_\sigma(\omega)} F_{\rho,k}(e^{z_\sigma(\omega)} u),$$

it follows then from (6.62a) and (6.62b) that

$$\begin{aligned} \|G_\rho(\omega, u)\| &\le Ce^{(k-1)z_\sigma(\omega)}\|u\|_\alpha^k, \quad &\forall\, u \in \mathscr{H}_\alpha,\ \omega \in \Omega, \\ \|G_\rho(\omega, u) - G_{\rho,k}(\omega, u)\| &\le Ce^{kz_\sigma(\omega)}\|u\|_\alpha^{k+1}, \quad &\forall\, u \in \mathscr{H}_\alpha,\ \omega \in \Omega. \end{aligned} \tag{6.65}$$

Similarly, it follows from (6.63) that

$$\begin{aligned} &\|G_{\rho,k}(\omega, u_1) - G_{\rho,k}(\omega, u_2)\| \\ &\qquad \le kCe^{(k-1)z_\sigma(\omega)}\big(\|u_1\|_\alpha^{k-1} + \|u_2\|_\alpha^{k-1}\big)\|u_1 - u_2\|_\alpha \\ &\qquad\quad + Ce^{kz_\sigma(\omega)}\|u_2\|_\alpha^k\|u_1 - u_2\|_\alpha, \quad \forall\, u_1, u_2 \in \mathscr{H}_\alpha,\ \omega \in \Omega. \end{aligned} \tag{6.66}$$

It follows directly from (6.65) and (6.66) that

$$\|P_{\mathfrak{s}}G_\rho(\theta_s\omega, u(s)) - P_{\mathfrak{s}}G_{\rho,k}(\theta_s\omega, u(s))\| \le Ce^{kz_\sigma(\theta_s\omega)}\|u(s)\|_\alpha^{k+1},\ \forall\, s \le 0, \tag{6.67}$$

and

$$\begin{aligned} &\|P_{\mathfrak{s}}G_{\rho,k}(\theta_s\omega, u(s)) - P_{\mathfrak{s}}G_{\rho,k}(\theta_s\omega, v(s))\| \\ &\qquad \le kCe^{(k-1)z_\sigma(\theta_s\omega)}\big(\|u(s)\|_\alpha^{k-1} + \|v(s)\|_\alpha^{k-1}\big)\|u(s) - v(s)\|_\alpha \\ &\qquad\quad + Ce^{kz_\sigma(\omega)}\|v(s)\|_\alpha^k\|u(s) - v(s)\|_\alpha, \quad \forall\, s \le 0. \end{aligned} \tag{6.68}$$

Now, we apply (6.60) and (6.67) and (6.68) together with the partial-dichotomy estimates given in (3.46) to estimate (6.59). For this purpose, let us choose η_1 and η_2 as in the proof of Proposition 6.1 with $2k$ in place of r. In particular,

$$0 > \eta_{\mathfrak{c}}^* > \eta_1 > \eta_2 > \eta_{\mathfrak{s}}^*, \tag{6.69}$$

recalling that $\eta_{\mathfrak{c}}^* = \eta_{\mathfrak{c}}(2k)$ and $\eta_{\mathfrak{s}}^* = \eta_{\mathfrak{s}}(2k)$.

The bounds in (6.67)–(6.68) allow us to control the terms in (6.60) which in turn allow us to control (6.59) after application of the partial-dichotomy estimate (3.46b):

$$\begin{aligned} &\|h_{\lambda,\rho}(\xi, \omega) - h_{\lambda,\rho}^{\mathrm{app}}(\xi, \omega)\|_\alpha \\ &\quad \le KC\int_{-\infty}^{0} \frac{e^{-\eta_2 s + \int_s^0 z_\sigma(\theta_\tau\omega)\,\mathrm{d}\tau + kz_\sigma(\theta_s\omega)}}{|s|^\alpha}\|u(s)\|_\alpha^{k+1}\mathrm{d}s \\ &\qquad + kKC\int_{-\infty}^{0}\Big(\frac{e^{-\eta_2 s + \int_s^0 z_\sigma(\theta_\tau\omega)\,\mathrm{d}\tau + (k-1)z_\sigma(\theta_s\omega)}}{|s|^\alpha} \\ &\qquad\qquad \times \big(\|u(s)\|_\alpha^{k-1} + \|v(s)\|_\alpha^{k-1}\big)\|u(s) - v(s)\|_\alpha\Big)\mathrm{d}s \end{aligned}$$

$$+ KC \int_{-\infty}^{0} \frac{e^{-\eta_2 s + \int_s^0 z_\sigma(\theta_\tau \omega)\,\mathrm{d}\tau + k z_\sigma(\theta_s \omega)}}{|s|^\alpha} \|v(s)\|_\alpha^k \|u(s) - v(s)\|_\alpha \mathrm{d}s \tag{6.70}$$

$$=: I_1(\xi, \omega) + I_2(\xi, \omega) + I_3(\xi, \omega).$$

To estimate the integral terms above, we first derive in the next step some estimates of $\|u(s)\|_\alpha$, $\|v(s)\|_\alpha$, and $\|u(s) - v(s)\|_\alpha$ for all $s \le 0$.

Step 3. Estimates about the stochastic flows $u(s)$ and $v(s)$. We show that for each $s \le 0$, $\xi \in \mathscr{H}^{\mathfrak{c}}$, and $\omega \in \Omega$, the following estimates hold:

$$\|u(s)\|_\alpha \le 2\,K e^{\eta s - \int_s^0 z_\sigma(\theta_\tau \omega)\,\mathrm{d}\tau} \|\xi\|_\alpha, \tag{6.71}$$

$$\|v(s)\|_\alpha \le K e^{\eta_1 s - \int_s^0 z_\sigma(\theta_\tau \omega)\,\mathrm{d}\tau} \|\xi\|_\alpha, \tag{6.72}$$

$$\|u(s) - v(s)\|_\alpha \le (2K)^k \mathfrak{C}_1(\omega) e^{(k\eta - (k-1)\varepsilon_1)s - \int_s^0 z_\sigma(\theta_\tau \omega)\,\mathrm{d}\tau} \|\xi\|_\alpha^k, \tag{6.73}$$

where $\mathfrak{C}_1(\omega)$ and ε_1 will be defined below in (6.85) and (6.79), respectively.

Recall that $u(\cdot) = u_{\lambda,\rho}(\cdot, \omega; \xi + h_{\lambda,\rho}(\xi, \omega))$ is the mild solution of Eq. (5.29) defined on $(-\infty, 0]$ taking value $\xi + h_{\lambda,\rho}(\xi, \omega)$ at $t = 0$, which is furthermore the unique fixed point in C_η^- of the operator $\mathscr{N}_{\xi,\rho}^{\omega,\lambda}$ defined in (6.54). In particular, $P_{\mathfrak{c}} u(0) = \xi$. It is also clear that $u(s) \equiv 0$ if $\xi = 0$. As a consequence, by using the estimate (B.16) applied to the mild solution $u(s)$ (with $\xi_1 = \xi$ and $\xi_2 = 0$), we obtain

$$\|u(\cdot)\|_{C_\eta^-} \le \frac{K}{1 - \Upsilon_1(F_\rho)} \|\xi\|_\alpha. \tag{6.74}$$

Recall from Theorem 5.1 that $\Upsilon_1(F_\rho)$ can be made less than $\frac{1}{2}$ by taking $\rho < \rho^*$. Using this in (6.74), we obtain for $\rho < \overline{\rho}$ (with $\overline{\rho}$ given in (6.64)):

$$\|u(\cdot)\|_{C_\eta^-} \le 2K \|\xi\|_\alpha. \tag{6.75}$$

Now, (6.71) follows directly from (6.75) and the definition of the $\|\cdot\|_{C_\eta^-}$-norm given in (4.4).

The estimate about $v(s)$ in (6.72) follows directly from the definition of v in (6.58) and the partial-dichotomy estimate (3.46c).

The remaining part of this step is devoted to deriving (6.73). Since $u(\cdot)$ is a fixed point in C_η^- of the integral operator $\mathscr{N}_{\xi,\rho}^{\omega,\lambda}$ defined in (6.54) and $v(\cdot) = \mathfrak{T}_{\lambda,\sigma}(\cdot, 0; \omega)\xi$, we get naturally:

$$
\begin{aligned}
u(s) - v(s) = &- \int_s^0 \mathfrak{T}_{\lambda,\sigma}(s, s'; \omega) P_{\mathfrak{c}} G_\rho(\theta_{s'}\omega, u(s'))\, \mathrm{d}s' \\
&+ \int_{-\infty}^s \mathfrak{T}_{\lambda,\sigma}(s, s'; \omega) P_{\mathfrak{s}} G_\rho(\theta_{s'}\omega, u(s'))\, \mathrm{d}s', \quad s \le 0,\ \omega \in \Omega.
\end{aligned} \tag{6.76}
$$

Applying the partial-dichotomy estimates (3.46) into (6.76), we obtain that

$$
\begin{aligned}
\|u(s) - v(s)\|_\alpha \le K &\int_s^0 e^{\eta_1(s-s') - \int_s^{s'} z_\sigma(\theta_\tau\omega)\,\mathrm{d}\tau} \|P_{\mathfrak{c}} G_\rho(\theta_{s'}\omega, u(s'))\|\, \mathrm{d}s' \\
&+ K \int_{-\infty}^s \frac{e^{\eta_2(s-s') + \int_{s'}^s z_\sigma(\theta_\tau\omega)\,\mathrm{d}\tau}}{|s-s'|^\alpha} \|P_{\mathfrak{s}} G_\rho(\theta_{s'}\omega, u(s'))\|\, \mathrm{d}s'.
\end{aligned}
$$

Using then (6.65) in the above estimate, we obtain for all $s \le 0$ and $\omega \in \Omega$ that

$$
\begin{aligned}
\|u(s) - v(s)\|_\alpha \le KC &\int_s^0 e^{\eta_1(s-s') - \int_s^{s'} z_\sigma(\theta_\tau\omega)\,\mathrm{d}\tau + (k-1) z_\sigma(\theta_{s'}\omega)} \|u(s')\|_\alpha^k\, \mathrm{d}s' \\
&+ KC \int_{-\infty}^s \frac{e^{\eta_2(s-s') + \int_{s'}^s z_\sigma(\theta_\tau\omega)\,\mathrm{d}\tau + (k-1) z_\sigma(\theta_{s'}\omega)}}{|s-s'|^\alpha} \|u(s')\|_\alpha^k\, \mathrm{d}s'.
\end{aligned} \tag{6.77}
$$

Before pursuing the analysis, we summarize in the following lemma some basic algebraic inequalities, which will be used to ensure mainly the existence of certain integrals arising in various places from the error estimates; see for instance (6.82), (6.88) and (6.89).

In that respect, let us recall that η_1 and η_2 chosen from the proof of Proposition 6.1 with $2k$ in place of r satisfy furthermore

$$
0 > 2k\eta_1 > \eta_2, \quad \text{and} \quad \eta \in \left(\frac{\eta_2}{2k}, \eta_1\right). \tag{6.78}
$$

Lemma 6.2 *Let ε_1 be the positive constant defined by:*

$$
\varepsilon_1 := \frac{2k\eta - \eta_2}{2(2k-1)}. \tag{6.79}
$$

Then, the following set of inequalities hold:

$$
\begin{aligned}
\eta_2 &< 2k\eta - (2k-1)\varepsilon_1 < (2k-1)\eta - 2(k-1)\varepsilon_1 \\
&\le (k+1)\eta - k\varepsilon_1 < k\eta - (k-1)\varepsilon_1 < \eta_1.
\end{aligned} \tag{6.80}
$$

Proof These inequalities can be checked directly using (6.69), (6.78), (6.79), and the fact that $k \geq 2$. For instance, the first inequality in (6.80) can be established as follows by using (6.79) and $2k\eta > \eta_2$:

$$2k\eta - (2k-1)\varepsilon_1 = 2k\eta - \frac{2k\eta - \eta_2}{2} = \frac{2k\eta + \eta_2}{2} > \eta_2.$$

To obtain the last inequality in (6.80), we note that

$$\eta_1 - (k\eta - (k-1)\varepsilon_1) = (\eta_1 - \eta) + (k-1)(\varepsilon_1 - \eta).$$

The desired result follows since $0 > \eta_1 > \eta$, $k - 1 > 0$, and $\varepsilon_1 > 0$. □

For $\varepsilon_1 > 0$ given in (6.79), let us introduce the following random variable:

$$C_{\varepsilon_1}(\omega) := \sup_{s' \leq 0} e^{\varepsilon_1 s' + |\int_{s'}^0 z_\sigma(\theta_\tau \omega)\,\mathrm{d}\tau| + |z_\sigma(\theta_{s'}\omega)|}. \tag{6.81}$$

Note that $C_{\varepsilon_1}(\omega)$ is finite for each $\omega \in \Omega$ thanks to the growth control estimates (3.30) of z_σ.

With the help of Lemma 6.2, the first integral on the RHS of (6.77) can be then controlled as follows:

$$\begin{aligned}
&\int_s^0 e^{\eta_1(s-s') - \int_s^{s'} z_\sigma(\theta_\tau\omega)\,\mathrm{d}\tau + (k-1)z_\sigma(\theta_{s'}\omega)} \|u(s')\|_\alpha^k \,\mathrm{d}s' \\
&\leq \int_s^0 e^{\eta_1(s-s') + k\eta s' - \int_s^0 z_\sigma(\theta_\tau\omega)\,\mathrm{d}\tau + (k-1)[z_\sigma(\theta_{s'}\omega) + \int_0^{s'} z_\sigma(\theta_\tau\omega)\,\mathrm{d}\tau]} \,\mathrm{d}s' \|u(\cdot)\|_{C_\eta^-}^k \\
&\leq C_{\varepsilon_1}^{k-1}(\omega) e^{-\int_s^0 z_\sigma(\theta_\tau\omega)\,\mathrm{d}\tau} \int_s^0 e^{\eta_1(s-s') + (k\eta - (k-1)\varepsilon_1)s'} \,\mathrm{d}s' \|u(\cdot)\|_{C_\eta^-}^k \\
&\leq \frac{e^{(k\eta - (k-1)\varepsilon_1)s - \int_s^0 z_\sigma(\theta_\tau\omega)\,\mathrm{d}\tau} C_{\varepsilon_1}^{k-1}(\omega)}{\eta_1 - k\eta + (k-1)\varepsilon_1} \|u(\cdot)\|_{C_\eta^-}^k, \quad \forall\, s \leq 0,
\end{aligned} \tag{6.82}$$

where the last integral above is well-defined since $\eta_1 - k\eta + (k-1)\varepsilon_1 > 0$ thanks to (6.80). Similarly,

$$\int_{-\infty}^{s} \frac{e^{\eta_2(s-s')+\int_{s'}^{s} z_\sigma(\theta_\tau\omega)\,d\tau+(k-1)z_\sigma(\theta_{s'}\omega)}}{|s-s'|^\alpha}\|u(s')\|_\alpha^k\,ds' \\ \le \frac{e^{(k\eta-(k-1)\varepsilon_1)s-\int_s^0 z_\sigma(\theta_\tau\omega)\,d\tau} C_{\varepsilon_1}^{k-1}(\omega)\Gamma(1-\alpha)}{(k\eta-\eta_2-(k-1)\varepsilon_1)^{1-\alpha}}\|u(\cdot)\|_{C_\eta^-}^k, \quad \forall s \le 0. \tag{6.83}$$

By using the controls obtained in (6.82) and (6.83), we deduce from (6.77) that

$$\|u(s)-v(s)\|_\alpha \le \mathfrak{C}_1(\omega)e^{(k\eta-(k-1)\varepsilon_1)s-\int_s^0 z_\sigma(\theta_\tau\omega)\,d\tau}\|u(\cdot)\|_{C_\eta^-}^k, \quad \forall s \le 0, \ \omega \in \Omega, \tag{6.84}$$

where

$$\mathfrak{C}_1(\omega) := KCC_{\varepsilon_1}^{k-1}(\omega)\Big(\frac{1}{\eta_1-k\eta+(k-1)\varepsilon_1}+\frac{\Gamma(1-\alpha)}{(k\eta-\eta_2-(k-1)\varepsilon_1)^{1-\alpha}}\Big). \tag{6.85}$$

Then, (6.73) follows from (6.84) by controlling $\|u(\cdot)\|_{C_\eta^-}$ using (6.75).

Step 4. Estimates of $\|h_{\lambda,\rho}(\xi,\omega)-h_{\lambda,\rho}^{\text{app}}(\xi,\omega)\|_\alpha$. Now we are ready to estimate $I_1(\xi,\omega)$, $I_2(\xi,\omega)$, and $I_3(\xi,\omega)$ as given in (6.70) in order to derive the estimate (6.57) at the end of this step.

Let us begin by estimating $I_1(\xi,\omega)$. Using (6.71) in $I_1(\xi,\omega)$, we obtain

$$\begin{aligned} I_1(\xi,\omega) &= KC\int_{-\infty}^{0} \frac{e^{-\eta_2 s+\int_s^0 z_\sigma(\theta_\tau\omega)\,d\tau+kz_\sigma(\theta_s\omega)}}{|s|^\alpha}\|u(s)\|_\alpha^{k+1}ds \\ &\le 2^{k+1}K^{k+2}C\int_{-\infty}^{0} \frac{e^{[(k+1)\eta-\eta_2]s-k\int_s^0 z_\sigma(\theta_\tau\omega)d\tau+kz_\sigma(\theta_s\omega)}}{|s|^\alpha}ds\,\|\xi\|_\alpha^{k+1} \\ &\le 2^{k+1}K^{k+2}CC_{\varepsilon_1}^k(\omega)\int_{-\infty}^{0} \frac{e^{[(k+1)\eta-\eta_2-k\varepsilon_1]s}}{|s|^\alpha}ds\,\|\xi\|_\alpha^{k+1} \\ &\le 2^{k+1}K^{k+2}CC_{\varepsilon_1}^k(\omega)\Gamma(1-\alpha)\big((k+1)\eta-\eta_2-k\varepsilon_1\big)^{\alpha-1}\|\xi\|_\alpha^{k+1} \\ &=: \mathfrak{C}_2(\omega)\|\xi\|_\alpha^{k+1}, \qquad \forall \xi \in \mathcal{H}^{\mathfrak{c}}, \ \omega \in \Omega, \end{aligned} \tag{6.86}$$

where the last integral above is well-defined since $(k+1)\eta-\eta_2-k\varepsilon_1 > 0$ thanks to (6.80).

For $I_2(\xi,\omega)$, first note that since $\eta < \eta_1 < 0$, by using the controls of $\|u(s)\|_\alpha$ and $\|v(s)\|_\alpha$ given respectively in (6.71) and (6.72) we obtain

$$\|u(s)\|_\alpha^{k-1}+\|v(s)\|_\alpha^{k-1} \le (2^{k-1}+1)K^{k-1}e^{(k-1)\eta s-(k-1)\int_s^0 z_\sigma(\theta_\tau\omega)d\tau}\|\xi\|_\alpha^{k-1}, \quad \forall s \le 0. \tag{6.87}$$

Now, $I_2(\xi,\omega)$ as defined in (6.70) can be controlled as follows by using (6.87) and (6.73):

$$\begin{aligned}
I_2(\xi,\omega) &\le 2^k(2^{k-1}+1)kK^{2k}C\mathfrak{C}_1(\omega) \\
&\quad\times \int_{-\infty}^{0} \frac{e^{[(2k-1)\eta-\eta_2-(k-1)\varepsilon_1]s+(k-1)[z_\sigma(\theta_s\omega)-\int_s^0 z_\sigma(\theta_\tau\omega)\,\mathrm{d}\tau]}}{|s|^\alpha}\mathrm{d}s\|\xi\|_\alpha^{2k-1} \\
&\le 2^k(2^{k-1}+1)kK^{2k}C\mathfrak{C}_1(\omega)C_{\varepsilon_1}^{k-1}(\omega)\int_{-\infty}^{0}\frac{e^{[(2k-1)\eta-\eta_2-2(k-1)\varepsilon_1]s}}{|s|^\alpha}\mathrm{d}s\|\xi\|_\alpha^{2k-1} \\
&= \frac{2^k(2^{k-1}+1)kK^{2k}C\mathfrak{C}_1(\omega)C_{\varepsilon_1}^{k-1}(\omega)\Gamma(1-\alpha)}{\big((2k-1)\eta-\eta_2-2(k-1)\varepsilon_1\big)^{1-\alpha}}\|\xi\|_\alpha^{2k-1} \\
&=: \mathfrak{C}_3(\omega)\|\xi\|_\alpha^{2k-1}, \qquad \forall\,\xi\in\mathcal{H}^{\mathrm{c}},\ \omega\in\Omega,
\end{aligned} \tag{6.88}$$

where the last integral above is well-defined since $(2k-1)\eta-\eta_2-2(k-1)\varepsilon_1>0$ thanks to (6.80).

By the same type of estimates, we obtain

$$\begin{aligned}
I_3(\xi,\omega) &= KC\int_{-\infty}^{0}\frac{e^{-\eta_2 s+\int_s^0 z_\sigma(\theta_\tau\omega)\,\mathrm{d}\tau+kz_\sigma(\theta_s\omega)}}{|s|^\alpha}\|v(s)\|_\alpha^k\|u(s)-v(s)\|_\alpha\mathrm{d}s \\
&\le \frac{2^kK^{2k+1}C\mathfrak{C}_1(\omega)C_{\varepsilon_1}^k(\omega)\Gamma(1-\alpha)}{\big(k\eta_1+k\eta-\eta_2-(2k-1)\varepsilon_1\big)^{1-\alpha}}\|\xi\|_\alpha^{2k} \\
&=: \mathfrak{C}_4(\omega)\|\xi\|_\alpha^{2k}, \qquad \forall\,\xi\in\mathcal{H}^{\mathrm{c}},\ \omega\in\Omega.
\end{aligned} \tag{6.89}$$

Note that $\mathfrak{C}_4(\omega)$ is positive since $k\eta_1+k\eta-\eta_2-(2k-1)\varepsilon_1>2k\eta-\eta_2-(2k-1)\varepsilon_1>0$ from (6.80).

Now, using the controls of $I_1(\xi,\omega)$ in (6.86), $I_2(\xi,\omega)$ in (6.88), and $I_3(\xi,\omega)$ in (6.89), we obtain then from (6.70) that

$$\begin{aligned}
&\|h_{\lambda,\rho}(\xi,\omega)-h_{\lambda,\rho}^{\mathrm{app}}(\xi,\omega)\|_\alpha \\
&\qquad\le \big(\mathfrak{C}_2(\omega)+\mathfrak{C}_3(\omega)\|\xi\|_\alpha^{k-2}+\mathfrak{C}_4(\omega)\|\xi\|_\alpha^{k-1}\big)\|\xi\|_\alpha^{k+1},
\end{aligned} \tag{6.90}$$

for all $\xi\in\mathcal{H}^{\mathrm{c}}$, $\lambda\in\Lambda_{2k}$, and all ω.

Let r_ε^* be now the positive random variable defined by

$$r_\varepsilon^*(\omega) := \min\left\{\frac{\varepsilon}{2\big(\mathfrak{C}_2(\omega)+\mathfrak{C}_3(\omega)+\mathfrak{C}_4(\omega)\big)},\ 1\right\}. \tag{6.91}$$

Then, we deduce for all $\xi\in B_\alpha(0,r_\varepsilon^*(\omega))\cap\mathcal{H}^{\mathrm{c}}$ and $\omega\in\Omega$ that

$$\Big(\mathfrak{C}_2(\omega) + \mathfrak{C}_3(\omega)\|\xi\|_\alpha^{k-2} + \mathfrak{C}_4(\omega)\|\xi\|_\alpha^{k-1}\Big)\|\xi\|_\alpha \leq \frac{\varepsilon}{2}.$$

This together with (6.90) implies that for each $\rho \in (0, \overline{\rho})$ with $\overline{\rho}$ given in (6.64), the following error estimate holds:

$$\begin{aligned} &\|h_{\lambda,\rho}(\xi,\omega) - h^{\mathrm{app}}_{\lambda,\rho}(\xi,\omega)\|_\alpha \\ &\qquad \leq \frac{\varepsilon}{2}\|\xi\|_\alpha^k, \ \ \forall\, \xi \in B_\alpha(0, r^*_\varepsilon(\omega)) \cap \mathcal{H}^{\mathfrak{c}},\ \lambda \in \Lambda_{2k},\ \omega \in \Omega, \end{aligned} \tag{6.92}$$

and (6.57) is proved.

Step 5. Estimates of $\|h^{\mathrm{app}}_\lambda(\xi,\omega) - h^{\mathrm{app}}_{\lambda,\rho}(\xi,\omega)\|_\alpha$. In this step, we get rid of the cut-off function from the expression of $h^{\mathrm{app}}_{\lambda,\rho}(\xi,\omega)$ given in (6.55), and provide a corresponding error estimate. The purpose here is to show that, for any $\rho \in (0,\overline{\rho})$ and any $\varepsilon > 0$, there exists a random open ball $B_\alpha(0, r^{**}_\varepsilon(\omega)) \subset \mathcal{H}_\alpha$ such that for all $\lambda \in \Lambda_{2k}$ the following estimate holds:

$$\|h^{\mathrm{app}}_\lambda(\xi,\omega) - h^{\mathrm{app}}_{\lambda,\rho}(\xi,\omega)\|_\alpha \leq \frac{\varepsilon}{2}\|\xi\|_\alpha^k, \ \ \forall \xi \in B_\alpha(0, r^{**}_\varepsilon(\omega)) \cap \mathcal{H}^{\mathfrak{c}},\ \omega \in \Omega, \tag{6.93}$$

where h^{app}_λ and $h^{\mathrm{app}}_{\lambda,\rho}$ are as defined in (6.28) and (6.55), respectively.

To do so, we first evaluate the difference $h^{\mathrm{app}}_\lambda - h^{\mathrm{app}}_{\lambda,\rho}$. In that respect, let us introduce

$$G_k(\omega,\ u) := e^{-z_\sigma(\omega)} F_k(e^{z_\sigma(\omega)}u) = e^{(k-1)z_\sigma(\omega)} F_k(u). \tag{6.94}$$

Then, by using (3.31), (6.94) and k-linear properties of F_k, the formula (6.28) can be rewritten as

$$h^{\mathrm{app}}_\lambda(\xi,\omega) = \int_{-\infty}^{0} \mathfrak{T}_{\lambda,\sigma}(0, s; \omega) P_{\mathfrak{s}} G_k(\theta_s\omega, \mathfrak{T}_{\lambda,\sigma}(s, 0; \omega)\xi)\, \mathrm{d}s. \tag{6.95}$$

Note that $G_{\rho,k}(\omega, u) = \zeta\Big(\frac{e^{z_\sigma(\omega)}\|u\|_\alpha}{\rho}\Big)G_k(\omega,\ u)$. Recall also $v(s) = \mathfrak{T}_{\lambda,\sigma}(s, 0; \omega)$ ξ is well-defined for $s < 0$ since $\xi \in \mathcal{H}^{\mathfrak{c}}$; see Sect. 3.4. It then follows from (6.55) and (6.95) that

$$\begin{aligned} &h^{\mathrm{app}}_\lambda(\xi,\omega) - h^{\mathrm{app}}_{\lambda,\rho}(\xi,\omega) \\ &= \int_{-\infty}^{0} \mathfrak{T}_{\lambda,\sigma}(0, s; \omega)\Big(1 - \zeta\Big(\frac{e^{z_\sigma(\theta_s\omega)}\|v(s)\|_\alpha}{\rho}\Big)\Big) P_{\mathfrak{s}} G_k(\theta_s\omega, v(s))\, \mathrm{d}s. \end{aligned} \tag{6.96}$$

The term $1 - \zeta\Big(\frac{e^{z_\sigma(\theta_s\omega)}\|v(s)\|_\alpha}{\rho}\Big)$ requires a special attention in order to control $\|h^{\mathrm{app}}_\lambda(\xi,\omega) - h^{\mathrm{app}}_{\lambda,\rho}(\xi,\omega)\|_\alpha$. From the definition of the cut-off function ζ given in (5.1), we have that

$$\left(e^{z_\sigma(\theta_s\omega)}\|v(s)\|_\alpha \le \rho\right) \Longrightarrow \left(\zeta\Big(\frac{e^{z_\sigma(\theta_s\omega)}\|v(s)\|_\alpha}{\rho}\Big) = 1\right). \tag{6.97}$$

Note furthermore that

$$e^{z_\sigma(\theta_s\omega)}\|v(s)\|_\alpha \le K e^{\eta_1 s - \int_s^0 z_\sigma(\theta_\tau\omega)\,\mathrm{d}\tau + z_\sigma(\theta_s\omega)} \|\xi\|_\alpha, \ \forall\, s \le 0,$$

which follows directly from (6.72). Hence,

$$e^{z_\sigma(\theta_s\omega)}\|v(s)\|_\alpha \le K C_{\varepsilon_1}(\omega) e^{(\eta_1-\varepsilon_1)s}\|\xi\|_\alpha, \ \forall\, s \le 0, \tag{6.98}$$

where $C_{\varepsilon_1}(\omega)$ is as defined in (6.81).

Recalling that $\eta_1 < 0$ thanks to (6.69), we then have $\eta_1 - \varepsilon_1 < 0$. This together with (6.98) implies that

$$\left(e^{z_\sigma(\theta_s\omega)}\|v(s)\|_\alpha > \rho\right) \Longrightarrow \left(s < \min\{0,\ s_0(\xi)\}\right), \tag{6.99}$$

where

$$s_0(\xi) := \frac{\log\big(\frac{\rho}{K C_{\varepsilon_1}(\omega)\|\xi\|_\alpha}\big)}{\eta_1 - \varepsilon_1}. \tag{6.100}$$

From now on, we consider ξ such that

$$\|\xi\|_\alpha < \frac{\rho}{K C_{\varepsilon_1}(\omega)}.$$

For such ξ, we have $s_0(\xi) < 0$. Then, it follows from (6.99) that $e^{z_\sigma(\theta_s\omega)}\|v(s)\|_\alpha \le \rho$ for all $s \in [s_0(\xi), 0]$ by contraposition. Hence, $\zeta\Big(\frac{e^{z_\sigma(\theta_s\omega)}\|v(s)\|_\alpha}{\rho}\Big) = 1$ for all $s \in [s_0(\xi), 0]$ according to (6.97).

We obtain thus from (6.96) the following control of $\|h_\lambda^{\mathrm{app}}(\xi,\omega) - h_{\lambda,\rho}^{\mathrm{app}}(\xi,\omega)\|_\alpha$:

$$\begin{aligned} &\|h_\lambda^{\mathrm{app}}(\xi,\omega) - h_{\lambda,\rho}^{\mathrm{app}}(\xi,\omega)\|_\alpha \\ &\le \Big\| \int_{-\infty}^{s_0(\xi)} \mathfrak{T}_{\lambda,\sigma}(0,s;\omega) P_{\mathfrak{s}} G_k(\theta_s\omega, v(s))\,\mathrm{d}s \Big\|_\alpha \\ &\le K \int_{-\infty}^{s_0(\xi)} \frac{e^{-\eta_2 s + \int_s^0 z_\sigma(\theta_\tau\omega)\,\mathrm{d}\tau}}{|s|^\alpha} \|P_{\mathfrak{s}} G_k(\theta_s\omega, v(s))\|\,\mathrm{d}s. \end{aligned} \tag{6.101}$$

The continuous k-linear property of F_k implies trivially the existence of $C > 0$ such that:

$$\|F_k(u)\| \le C\|u\|_\alpha^k, \qquad \forall\, u \in \mathcal{H}_\alpha,$$

leading thus to

$$\begin{aligned}\|G_k(\theta_s\omega, v(s))\| &= e^{-z_\sigma(\theta_s\omega)}\|F_k(e^{z_\sigma(\theta_s\omega)}v(s))\| \le Ce^{(k-1)z_\sigma(\theta_s\omega)}\|v(s)\|_\alpha^k \\ &\le CK^k e^{k\eta_1 s+(k-1)z_\sigma(\theta_s\omega)-k\int_s^0 z_\sigma(\theta_\tau\omega)\,\mathrm{d}\tau}\|\xi\|_\alpha^k, \quad \forall\, s \le 0,\end{aligned}$$

where the control of $\|v(s)\|_\alpha$ given in (6.72) is used to derive the last inequality above.

We obtain then from (6.101):

$$\begin{aligned}&\|h_\lambda^{\mathrm{app}}(\xi,\omega) - h_{\lambda,\rho}^{\mathrm{app}}(\xi,\omega)\|_\alpha \\ &\le CK^{k+1}\int_{-\infty}^{s_0(\xi)} \frac{e^{(k\eta_1-\eta_2)s+(k-1)[z_\sigma(\theta_s\omega)-\int_s^0 z_\sigma(\theta_\tau\omega)\,\mathrm{d}\tau]}}{|s|^\alpha}\,\mathrm{d}s\,\|\xi\|_\alpha^k \\ &=: J(\omega)\|\xi\|_\alpha^k.\end{aligned} \tag{6.102}$$

We conclude now about the control of $\|h_\lambda^{\mathrm{app}}(\xi,\omega) - h_{\lambda,\rho}^{\mathrm{app}}(\xi,\omega)\|_\alpha$ by showing that the random constant $J(\omega)$ given in (6.102) can be actually dominated by $\varepsilon/2$ provided that $\|\xi\|_\alpha$ is sufficiently small. This control relies on the fact that $s_0(\xi)$ as defined in (6.100) becomes more negative as $\|\xi\|_\alpha$ gets smaller.

By noting that

$$\begin{aligned}&(k\eta_1-\eta_2)s + (k-1)\Big[z_\sigma(\theta_s\omega) - \int_s^0 z_\sigma(\theta_\tau\omega)\,\mathrm{d}\tau\Big] \\ &\le (k\eta_1-\eta_2)s - (k-1)\varepsilon_1 s \\ &\quad + (k-1)\Big(\varepsilon_1 s + |z_\sigma(\theta_s\omega)| + \Big|\int_s^0 z_\sigma(\theta_\tau\omega)\,\mathrm{d}\tau\Big|\Big),\end{aligned}$$

the constant $C_{\varepsilon_1}(\omega)$ given in (6.81) appears in the control of $J(\omega)$ as follows:

$$\begin{aligned}J(\omega) &\le \frac{CK^{k+1}C_{\varepsilon_1}^{k-1}(\omega)}{|s_0(\xi)|^\alpha}\int_{-\infty}^{s_0(\xi)} e^{(k\eta_1-\eta_2-(k-1)\varepsilon_1)s}\,\mathrm{d}s \\ &= \frac{CK^{k+1}C_{\varepsilon_1}^{k-1}(\omega)}{(k\eta_1-\eta_2-(k-1)\varepsilon_1)|s_0(\xi)|^\alpha}e^{(k\eta_1-\eta_2-(k-1)\varepsilon_1)s_0(\xi)},\end{aligned}$$

where $k\eta_1 - \eta_2 - (k-1)\varepsilon_1 > k\eta - \eta_2 - (k-1)\varepsilon_1 > 0$ thanks to (6.80).

Now by recalling the definition of $s_0(\xi)$ in (6.100), we obtain then:

$$J(\omega) \leq \frac{CK^{k+1}C_{\varepsilon_1}^{k-1}(\omega)|\eta_1 - \varepsilon_1|^{\alpha}}{(k\eta_1 - \eta_2 - (k-1)\varepsilon_1)} \frac{1}{|\log(\frac{\rho}{KC_{\varepsilon_1}(\omega)\|\xi\|_\alpha})|^{\alpha}} \Big(\frac{\rho}{KC_{\varepsilon_1}(\omega)\|\xi\|_\alpha}\Big)^{\eta_\varepsilon},$$

with

$$\eta_\varepsilon := \frac{k\eta_1 - \eta_2 - (k-1)\varepsilon_1}{\eta_1 - \varepsilon_1}.$$

Let us introduce furthermore the following positive random variable

$$\widetilde{r}_\varepsilon(\omega) := \frac{\rho}{KC_{\varepsilon_1}(\omega)} \Big(\frac{\varepsilon[k\eta_1 - \eta_2 - (k-1)\varepsilon_1]}{2CK^{k+1}C_{\varepsilon_1}^{k-1}(\omega)|\eta_1 - \varepsilon_1|^{\alpha}}\Big)^{-\frac{1}{\eta_\varepsilon}}.$$

We obtain then for any $\xi \in \mathscr{H}^{\mathfrak{c}}$ such that $\|\xi\|_\alpha < \widetilde{r}_\varepsilon(\omega)$:

$$\Big(\frac{\rho}{KC_{\varepsilon_1}(\omega)\|\xi\|_\alpha}\Big)^{\eta_\varepsilon} < \Big(\frac{\rho}{KC_{\varepsilon_1}(\omega)\widetilde{r}_\varepsilon(\omega)}\Big)^{\eta_\varepsilon} = \frac{\varepsilon[k\eta_1 - \eta_2 - (k-1)\varepsilon_1]}{2CK^{k+1}C_{\varepsilon_1}^{k-1}(\omega)|\eta_1 - \varepsilon_1|^{\alpha}},$$

since $\eta_\varepsilon < 0$ due to the fact that $\eta_1 < 0$ from (6.69), and $k\eta_1 - \eta_2 - (k-1)\varepsilon_1 > 0$ from what precedes.

As a consequence,

$$\frac{CK^{k+1}C_{\varepsilon_1}^{k-1}(\omega)|\eta_1 - \varepsilon_1|^{\alpha}}{(k\eta_1 - \eta_2 - (k-1)\varepsilon_1)} \Big(\frac{\rho}{KC_{\varepsilon_1}(\omega)\|\xi\|_\alpha}\Big)^{\eta_\varepsilon} < \frac{\varepsilon}{2}, \quad \forall \xi \in B_\alpha(0, \widetilde{r}_\varepsilon(\omega)) \cap \mathscr{H}^{\mathfrak{c}}.$$

Note also that

$$\log\Big(\frac{\rho}{KC_{\varepsilon_1}(\omega)\|\xi\|_\alpha}\Big) > 1, \quad \forall \xi \in B_\alpha\Big(0, \frac{\rho}{eKC_{\varepsilon_1}(\omega)}\Big) \cap \mathscr{H}^{\mathfrak{c}},$$

where e denotes the Euler constant.

Now, by introducing

$$r_\varepsilon^{**}(\omega) := \min\Big\{\frac{\rho}{eKC_{\varepsilon_1}(\omega)}, \ \widetilde{r}_\varepsilon(\omega)\Big\}, \tag{6.103}$$

we have that

$$J(\omega) < \frac{\varepsilon}{2}, \quad \forall \xi \in B(0, r_\varepsilon^{**}(\omega)), \ \omega \in \Omega. \tag{6.104}$$

The desired control (6.93) on $\|h_\lambda^{\mathrm{app}}(\xi,\omega) - h_{\lambda,\rho}^{\mathrm{app}}(\xi,\omega)\|_\alpha$ is thus achieved from (6.102).

Step 6. Estimates of $\|h_\lambda(\xi,\omega) - h_\lambda^{\text{app}}(\xi,\omega)\|_\alpha$. Let

$$r_\varepsilon := \min\left\{r_\varepsilon^*, r_\varepsilon^{**}\right\}, \tag{6.105}$$

where r_ε^* and r_ε^{**} are given by (6.91) and (6.103), respectively.

It follows then from (6.57) and (6.93) that for each $\rho \in (0, \overline{\rho})$:

$$\|h_{\lambda,\rho}(\xi,\omega) - h_\lambda^{\text{app}}(\xi,\omega)\|_\alpha \le \varepsilon \|\xi\|_\alpha^k, \; \forall \xi \in B_\alpha(0, r_\varepsilon(\omega)) \cap \mathcal{H}^{\mathfrak{c}}, \; \lambda \in \Lambda_{2k}, \; \omega \in \Omega. \tag{6.106}$$

The estimate in (6.29) follows then from (6.106) by recalling from Step 1 that $h_\lambda = h_{\lambda,\rho}$.

By recalling that $\widehat{h}_\lambda(\xi,\omega) = e^{z_\sigma(\omega)} h_\lambda(e^{-z_\sigma(\omega)}\xi, \omega)$ from (5.13), the corresponding error estimate for $\widehat{h}_\lambda^{\text{app}}$ given in (6.31) follows then directly from (6.29). The proof is complete. □

Proof of Corollary 6.1 Since the eigenvectors of L_λ form a Hilbert basis of $\mathcal{H}$, the approximation h_λ^{app} defined in (6.28) can be expanded as given in (6.36) with the random coefficients $h_\lambda^{\text{app},n}$ determined for all $n \ge m+1$ as follows:

$$\begin{aligned} &h_\lambda^{\text{app},n}(\xi,\omega) \\ &\quad = \langle h_\lambda^{\text{app}}(\xi,\omega), e_n\rangle \\ &\quad = e^{(k-1)z_\sigma(\omega)} \int_{-\infty}^{0} \left\langle e^{\sigma(k-1)W_s(\omega)\text{Id}} e^{-sL_\lambda} P_{\mathfrak{s}} F_k\left(e^{sL_\lambda}\xi\right), e_n\right\rangle \mathrm{d}s, \quad \xi \in \mathcal{H}^{\mathfrak{c}}, \; \omega \in \Omega. \end{aligned} \tag{6.107}$$

We check now that $h_\lambda^{\text{app},n}$ can be written in the form as given by (6.37).

First note that, for any $\xi \in \mathcal{H}^{\mathfrak{c}}$, it can be written as $\xi = \sum_{i=1}^m \xi_i e_i$, with $\xi_i = \langle \xi, e_i\rangle$, $i = 1, \ldots, m$. Then,

$$e^{sL_\lambda}\xi = \sum_{i=1}^m \xi_i e^{sL_\lambda} e_i = \sum_{i=1}^m e^{\beta_i(\lambda)s}\xi_i e_i, \quad \forall\, \xi \in \mathcal{H}^{\mathfrak{c}}, \; s \le 0. \tag{6.108}$$

Note also that for any $s \le 0$, $u \in \mathcal{H}$, and $n > m$, we have

$$\left\langle e^{-sL_\lambda} P_{\mathfrak{s}} u, e_n\right\rangle = \left\langle P_{\mathfrak{s}} u, e^{-sL_\lambda^*} e_n\right\rangle = \left\langle P_{\mathfrak{s}} u, e^{-s\beta_n(\lambda)} e_n\right\rangle = e^{-s\beta_n(\lambda)}\left\langle u, e_n\right\rangle, \tag{6.109}$$

where in the last equality above we used $\langle P_{\mathfrak{c}} u, e_n\rangle = 0$ since $n > m$.

Now, using the identities (6.108) and (6.109) in (6.107) we obtain

$$
\begin{aligned}
&h_\lambda^{\mathrm{app},n}(\xi, \omega) \\
&= e^{(k-1)z_\sigma(\omega)} \int_{-\infty}^{0} \Big\langle e^{\sigma(k-1)W_s(\omega)\mathrm{Id}} e^{-sL_\lambda} P_{\mathfrak{s}} F_k\Big(\sum_{i=1}^{m} e^{\beta_i(\lambda)s}\xi_i e_i\Big), e_n\Big\rangle \mathrm{d}s \\
&= e^{(k-1)z_\sigma(\omega)} \int_{-\infty}^{0} \Big\langle e^{\sigma(k-1)W_s(\omega)-s\beta_n(\lambda)} F_k\Big(\sum_{i=1}^{m} e^{\beta_i(\lambda)s}\xi_i e_i\Big), e_n\Big\rangle \mathrm{d}s.
\end{aligned} \tag{6.110}
$$

Since F_k is k-linear, we have

$$
F_k\Big(\sum_{i=1}^{m} e^{\beta_i(\lambda)s}\xi_i e_i\Big) = \sum_{(i_1,\ldots,i_k)\in\mathscr{I}^k} \xi_{i_1},\ldots,\xi_{i_k} F_k(e_{i_1},\ldots,e_{i_k}) e^{\sum_{j=1}^{k}\beta_{i_j}(\lambda)s},
$$

where $\mathscr{I} = \{1, \ldots, m\}$. We get then

$$
\begin{aligned}
&h_\lambda^{\mathrm{app},n}(\xi, \omega) \\
&= e^{(k-1)z_\sigma(\omega)} \sum_{(i_1,\ldots,i_k)\in\mathscr{I}^k} \xi_{i_1},\ldots,\xi_{i_k} \langle F_k(e_{i_1},\ldots,e_{i_k}), e_n\rangle \\
&\quad\times \int_{-\infty}^{0} e^{\left(\sum_{j=1}^{k}\beta_{i_j}(\lambda)-\beta_n(\lambda)\right)s+\sigma(k-1)W_s(\omega)} \mathrm{d}s \\
&= e^{(k-1)z_\sigma(\omega)} \sum_{(i_1,\ldots,i_k)\in\mathscr{I}^k} \xi_{i_1}\ldots\xi_{i_k} \langle F_k(e_{i_1},\ldots,e_{i_k}), e_n\rangle M_n^{i_1,\ldots,i_k}(\omega,\lambda),
\end{aligned} \tag{6.111}
$$

where

$$
M_n^{i_1,\ldots,i_k}(\omega,\lambda) := \int_{-\infty}^{0} e^{\left(\sum_{j=1}^{k}\beta_{i_j}(\lambda)-\beta_n(\lambda)\right)s+\sigma(k-1)W_s(\omega)} \mathrm{d}s.
$$

The formula (6.37) is now obtained.

It is clear from (6.111) that $h_\lambda^{\mathrm{app},n}(\xi,\cdot)$ is $\big(\mathscr{F};\mathscr{B}(\mathbb{R})\big)$-measurable for each fixed $\xi \in \mathscr{H}^{\mathrm{c}}$, and $h_\lambda^{\mathrm{app},n}(\cdot,\omega)$ is a homogeneous polynomial in $\xi_1,\ldots,\xi_m$ of order k for each ω. Hence, h_λ^{app} is a random homogeneous polynomial of order k in the sense given in Definition 6.2. This together with the estimate (6.29) implies that h_λ^{app} constitutes the leading-order Taylor approximation of h_λ.

The corresponding results for $\widehat{h}_\lambda^{\mathrm{app}}$ given in (6.39) and (6.40) can be derived using the relation $\widehat{h}_\lambda(\xi,\omega) = e^{z_\sigma(\omega)}h_\lambda(e^{-z_\sigma(\omega)}\xi,\omega)$; see (5.13). The proof is now complete. □

Chapter 7
Approximation of Stochastic Hyperbolic Invariant Manifolds

The approximation formulas provided by Theorem 6.1 and Corollary 6.1 were presented in the case where the subspace $\mathscr{H}^{\mathfrak{c}}$ contains only critical modes which lose their stability, formulated in terms of the PES condition (6.4), as the control parameter λ varies; see also Remark 6.1. In practice, it can be also of interest to consider other situations where the subspace $\mathscr{H}^{\mathfrak{c}}$ contain a combination of critical modes and modes that remain stable as λ varies in some interval Λ. It is then natural to ask whether the formulas provided by Theorem 6.1 and Corollary 6.1 still provide approximation to the leading order of the corresponding local stochastic invariant manifolds.[1]

First, note that by adding stable modes into $\mathscr{H}^{\mathfrak{c}}$ which remain stable as λ varies, then when $\lambda > \lambda_c$ the subspace $\mathscr{H}^{\mathfrak{c}}$ is spanned by a mixture of stable and unstable modes, making thus *hyperbolic* the corresponding local stochastic invariant manifold. This property results from the fact that $\mathscr{H}^{\mathfrak{c}}$—which is tangent to the local stochastic manifold at the origin—is then decomposed as the direct sum of the unstable subspace and the subspace spanned by the stable modes contained in $\mathscr{H}^{\mathfrak{c}}$. We are thus concerned here with the approximation of such hyperbolic stochastic invariant manifolds, which is the content of the following corollary. Here, the PES condition required in Theorem 6.1 and Corollary 6.1 is replaced by the condition (7.1) below.

Corollary 7.1 *Consider the SPDE* (3.1). *Assume that all the assumptions given in Sect.* 6.1 *are fulfilled except the PES condition* (6.4). *We assume furthermore that an open interval* Λ *is chosen such that the uniform spectrum decomposition as given in* (3.11) *holds over* Λ, *and that there exist* η_1 *and* η_2 *such that*

$$\eta_{\mathfrak{s}} < \eta_2 < \eta_1 < \eta_{\mathfrak{c}}, \qquad \eta_2 < 2k\eta_1 < 0, \tag{7.1}$$

where $k \geq 2$ *denotes the leading order of the nonlinear term* F*; see* (6.1). *Let* $\mathscr{H}^{\mathfrak{c}}$ *and* $\mathscr{H}^{\mathfrak{s}}_{\alpha}$ *be the corresponding subspaces associated with the uniform spectrum decomposition as defined in* (3.18) *and* (3.20) *with* $\dim(\mathscr{H}^{\mathfrak{c}}) = m$.

[1] According to Corollary 5.1, the latter always exist in a sufficiently small neighborhood of the origin.

M.D. Chekroun et al., *Approximation of Stochastic Invariant Manifolds*,
SpringerBriefs in Mathematics, DOI 10.1007/978-3-319-12496-4_7

Then, for each $\eta \in (\frac{\eta_2}{2k}, \eta_1)$*, there exists* $\rho^* > 0$*, such that for each* $\rho \in (0, \rho^*)$*, Eq.* (3.1) *possesses a family of local stochastic invariant* C^1*-manifolds* $\{\widehat{\mathfrak{M}}_\lambda^{\text{loc}}\}_{\lambda \in \Lambda}$*, where each of such manifolds is m-dimensional and takes the abstract form given by* (5.10).

Moreover, the corresponding local stochastic manifold function $\widehat{h}_\lambda$ *is approximated*[2] *to the leading order, k, by* $\widehat{h}_\lambda^{\text{app}}$ *as defined in* (AF)*, that is*

$$\boxed{\widehat{h}_\lambda^{\text{app}}(\xi, \omega) = \int_{-\infty}^{0} e^{\sigma(k-1)W_s(\omega)\text{Id}} e^{-sL_\lambda} P_{\mathfrak{s}} F_k(e^{sL_\lambda}\xi)\, \text{d}s, \quad \forall\, \xi \in \mathcal{H}^{\mathfrak{c}}, \omega \in \Omega.} \tag{7.2}$$

If the linear operator L_λ *is furthermore assumed to be self-adjoint, then the approximation* $\widehat{h}_\lambda^{\text{app}}$ *constitutes the leading-order Taylor approximation of* $\widehat{h}_\lambda$*, which can be rewritten into the form given by* (6.39) *and* (6.40).

Remark 7.1 Note that the existence of ρ^* in the above theorem is related to the uniform spectral gap conditions (5.5) and (5.6), which are subordinated to the choice of η in $(\frac{\eta_2}{2k}, \eta_1)$.

Proof The existence of local stochastic invariant C^1-manifolds follows directly from Corollary 5.1. The proof consists then of noting that in the derivation of Theorem 6.1, the PES condition (6.4) is only used to ensure—via Proposition 6.1—that the condition $\eta_2 < 2k\eta_1 < 0$ holds in order to apply the technical Lemma 6.2 for the control of certain integrals as pointed out in the description of the skeleton of the proof of Theorem 6.1; see also (6.78).

The assumption (7.1) allows us to get rid of the PES condition, and the approximation results can thus be derived by following the same lines of proofs of Theorem 6.1 and Corollary 6.1. □

As just explained in the proof above, the condition $\eta_2 < 2k\eta_1 < 0$ of Corollary 7.1 is required here for purely technical reasons in order to guarantee certain integrals emerging from the estimates to converge; see, e.g., Lemma 6.2 and Step 4 of the proof of Theorem 6.1.

When this condition is not satisfied, it is reasonable to conjecture that the Lyapunov-Perron integral $\mathfrak{I}_\lambda$ given in (6.25) still gives, when it exists, the leading-order approximation of $\widehat{\mathfrak{M}}_\lambda^{\text{loc}}$. In the general case we saw in Sect. 6.2 that the condition $\eta_2 < k\eta_1$ was sufficient to ensure the existence of $\mathfrak{I}_\lambda$ when λ varies in some interval Λ, so that the conclusions of Corollary 7.1 should still hold under this weaker condition. In the case where L_λ is self-adjoint, a necessary and sufficient condition for $\mathfrak{I}_\lambda$ to exist can be formulated as a *cross non-resonance condition* given in (NR) below.

It is illustrated in Volume II [37, Chaps. 6–7] in the case of a stochastic Burgers-type equation, that when the (NR)-condition is satisfied, the formula (6.39) given in Corollary 6.1 provides still an efficient tool to derive reduced equations for the

[2] In the sense given by (6.33).

amplitudes of the modes in $\mathscr{H}^{\mathfrak{c}}$, in the case where stable modes are included in $\mathscr{H}^{\mathfrak{c}}$ while the condition $\eta_2 < 2k\eta_1 < 0$ is violated.

The aforementioned cross non-resonance condition for the self-adjoint case can be described as follows. If the leading-order nonlinear interactions between the low modes in $\mathscr{H}^{\mathfrak{c}}$ when projected against a given high mode is not zero, then the corresponding eigenvalues associated with these low modes and the given high mode should satisfy a cross non-resonance condition of the following form[3]:

$$\boxed{\begin{aligned}&\forall\ (i_1, \ldots, i_k) \in \mathscr{I}^k,\ n > m,\ \lambda \in \Lambda,\ \text{it holds that}\\ &\Big(\langle F_k(e_{i_1}, \ldots, e_{i_k}), e_n\rangle \neq 0\Big) \Longrightarrow \Big(\sum_{j=1}^{k} \beta_{i_j}(\lambda) - \beta_n(\lambda) > 0\Big),\end{aligned}} \tag{NR}$$

where m is the dimension of $\mathscr{H}^{\mathfrak{c}}$ spanned by the first m eigenvectors of L_λ, i.e.,

$$\mathscr{H}^{\mathfrak{c}} := \text{span}\{e_1, \ldots, e_m\},$$

and where $\mathscr{I} = \{1, \ldots, m\}$, and $\langle\cdot,\cdot\rangle$ denotes the inner-product in the ambient Hilbert space $\mathscr{H}$. As for Corollary 6.1, each $\beta_{i_j}(\lambda)$ denotes the eigenvalue associated with the corresponding mode e_{i_j} in $\mathscr{H}^{\mathfrak{c}}$, and $\beta_n(\lambda)$ denotes the eigenvalue associated with the mode e_n in $\mathscr{H}^{\mathfrak{s}}$. In fact, the (NR)-condition is needed for the corresponding $M_n^{i_1,\ldots,i_k}$-terms, given in (6.38) of Corollary 6.1, to be finite so that the expression of $\widehat{h}_\lambda^{\text{app},n}$ given in (6.40) is well-defined.

Remark 7.2 We mention that cross non-resonances similar to (NR) arise in the power series expansion of invariant manifolds, and normal forms of deterministic (and finite-dimensional) vector fields near an equilibrium; see, e.g., [14, Theorem 3.1] and [91].

As illustrated in Volume II [37], when the (NR)-condition is met, the formula (6.39) characterizes in certain cases a (hyperbolic) manifold which does not necessarily approximate an invariant manifold, but still conveys very useful dynamical information; see [37, Chaps. 6–7]. This is formulated via the concept of *stochastic parameterizing manifolds* introduced in [37, Sect. 4.2], for which it is shown that their construction can be performed by means of pullback limits such as described in Volume II [37].

Example We provide below an illustration of Corollary 7.1. For that purpose, we consider on the bounded interval $(0, l)$ with Dirichlet boundary conditions, the following SPDE

$$du = \big(\nu u_{xx} + \lambda u - \gamma u u_x\big)dt + \sigma u \circ dW_t, \tag{7.3}$$

i.e. a stochastic Burger-type equation for which we refer the reader to [37, Chap. 6] for more details.

[3] See [4, Sect. 22 A] for the definition of a more standard notion of non-resonance.

The goal is here to show that approximation formulas such as (7.2) can be efficiently computed, and can furthermore allow us to get access to precise geometric information about the corresponding (stochastic) approximating hyperbolic manifolds.

In that perspective, we consider

$$\mathcal{H}^{\mathfrak{c}} := \text{span}\{e_1, e_2\},$$

where $e_j(x) := \sqrt{\frac{2}{l}} \sin\left(\frac{j\pi}{l}x\right)$ ($j \in \{1, 2\}$) correspond to the first two eigenvectors associated with the linear part of Eq. (7.3). By recalling that the corresponding eigenvalues are given by

$$\beta_j(\lambda) = \lambda - \frac{\nu j^2 \pi^2}{l^2}, \tag{7.4}$$

it turns out that $\beta_2(\lambda) < 0$ and $\beta_1(\lambda) > 0$, as λ varies in $\Lambda := (\frac{\nu\pi^2}{l^2}, \frac{4\nu\pi^2}{l^2})$.

By noting that

$$\begin{aligned} &\langle e_1(e_2)_x, e_3\rangle = \frac{\sqrt{2}\pi}{l^{3/2}}, \quad \langle e_2(e_1)_x, e_3\rangle = \frac{\pi}{\sqrt{2}l^{3/2}}, \quad \langle e_2(e_2)_x, e_4\rangle = \frac{\sqrt{2}\pi}{l^{3/2}}, \\ &\langle e_i(e_j)_x, e_n\rangle = 0, \qquad i, j \in \{1,\ 2\},\ n \geq 5, \end{aligned} \tag{7.5}$$

where $\langle\cdot,\cdot\rangle$ denotes the L^2-inner product, the (NR)-condition reduces here to:

$$\beta_1(\lambda) + \beta_2(\lambda) - \beta_3(\lambda) > 0, \qquad 2\beta_2(\lambda) - \beta_4(\lambda) > 0, \tag{7.6}$$

which can be trivially checked to be satisfied for $\lambda \in \Lambda$.

Note also that for the chosen resolved subspace $\mathcal{H}^{\mathfrak{c}}$ and the given interval Λ, it holds that $\eta_{\mathfrak{c}} = -\frac{3\nu\pi^2}{l^2}$ and $\eta_{\mathfrak{s}} = -\frac{5\nu\pi^2}{l^2}$; cf. (3.12). We can then apply Corollary 5.1 to conclude the existence of a family of local stochastic invariant C^1-manifolds $\widehat{\mathfrak{M}}_\lambda^{\text{loc}}$ which are hyperbolic.

Now, by application of (7.2) in the self-adjoint case (see (6.39) and (6.40)), we obtain that $\widehat{\mathfrak{M}}_\lambda^{\text{loc}}$ is approximated by

$$\widehat{\mathfrak{M}}_\lambda^{\text{app}}(\omega) = \left\{y_1 e_1 + y_2 e_2 + \widehat{h}_\lambda^{\text{app}}(y_1 e_1 + y_2 e_2, \omega) : (y_1, y_2) \in \Gamma(\omega)\right\}, \quad \omega \in \Omega, \tag{7.7}$$

where $\Gamma(\omega)$ is some random neighborhood of the origin, and $\widehat{h}_\lambda^{\text{app}}$ is given by

$$\begin{aligned} \widehat{h}_\lambda^{\text{app}}(y_1 e_1 + y_2 e_2, \omega) = &-\frac{3\gamma\pi}{\sqrt{2}l^{3/2}} M_3^{12}(\omega, \lambda) y_1 y_2 e_3 \\ &- \frac{\sqrt{2}\gamma\pi}{l^{3/2}} M_4^{22}(\omega, \lambda) y_2^2 e_4, \end{aligned} \tag{7.8}$$

with the random coefficients $M_3^{12}(\omega, \lambda)$ and $M_4^{22}(\omega, \lambda)$ given by

$$\begin{aligned} M_3^{12}(\omega,\lambda) &= \int_{-\infty}^{0} e^{[\beta_1(\lambda)+\beta_2(\lambda)-\beta_3(\lambda)]s+\sigma W_s(\omega)}\,\mathrm{d}s, \\ M_4^{22}(\omega,\lambda) &= \int_{-\infty}^{0} e^{[2\beta_2(\lambda)-\beta_4(\lambda)]s+\sigma W_s(\omega)}\,\mathrm{d}s. \end{aligned} \tag{7.9}$$

The approximating manifold $\widehat{\mathfrak{M}}_\lambda^{\text{app}}$ can then be represented as

$$\widehat{\mathfrak{M}}_\lambda^{\text{app}}(\omega) := \bigcup_{x\in[0,l]} \widehat{\mathfrak{M}}_{\lambda,x}^{\text{app}}(\omega), \tag{7.10}$$

where for each $x \in [0, l]$, the x-section,[4] $\widehat{\mathfrak{M}}_{\lambda,x}^{\text{app}}$, is given by:

$$\widehat{\mathfrak{M}}_{\lambda,x}^{\text{app}}(\omega) := \Big\{\big(y_1e_1(x), y_2e_2(x), f_\lambda^x(y_1, y_2; \omega)\big) : (y_1, y_2) \in \Gamma(\omega)\Big\}, \quad \omega \in \Omega, \tag{7.11}$$

with

$$\begin{aligned} f_\lambda^x(y_1, y_2; \omega) := &-\frac{3\gamma\pi}{\sqrt{2}l^{3/2}} M_3^{12}(\omega,\lambda)y_1y_2e_3(x) \\ &- \frac{\sqrt{2}\gamma\pi}{l^{3/2}} M_4^{22}(\omega,\lambda)y_2^2e_4(x). \end{aligned} \tag{7.12}$$

It is the explicit representation of $\widehat{\mathfrak{M}}_\lambda^{\text{app}}$ provided by (7.10)–(7.12) that allowed us to produce the opening figure of this monograph. This figure displays different x-sections at various values of x and for a given realization ω. To do so, the random coefficients M_3^{12} and M_4^{22} are not obtained from their integral expressions given in (7.9), but rather from numerical integration of the following auxiliary SDEs

$$\mathrm{d}M = \Big(1 - \big(\beta_1(\lambda) + \beta_2(\lambda) - \beta_3(\lambda)\big)M\Big)\,\mathrm{d}t - \sigma M \circ \mathrm{d}W_t, \tag{7.13}$$

and

$$\mathrm{d}M = \Big(1 - \big(2\beta_2(\lambda) - \beta_4(\lambda)\big)M\Big)\,\mathrm{d}t - \sigma M \circ \mathrm{d}W_t, \tag{7.14}$$

see [37, Sect. 5.3] for more details.

As a byproduct, the analytic expression (7.12) combined with the approximation result provided by Corollary 7.1, allow us to provide useful geometric information of the hyperbolic stochastic invariant $\widehat{\mathfrak{M}}_\lambda^{\text{loc}}$, near the origin. For instance, given a

[4]Called x-disintegration in the caption of the opening figure of this monograph.

realization ω, the Gaussian curvature[5] at the origin, of its x-section, $\widehat{\mathfrak{M}}^{\mathrm{loc}}_{\lambda,x}$, is approximated by

$$K_x(\omega) := -\frac{9\gamma^2\pi^2(M_3^{12}(\omega,\lambda))^2(e_3(x))^2}{2l^3(e_1(x)e_2(x))^2}, \quad x \notin \{0,\ l/3,\ l/2,\ 2l/3,\ l\}.$$

[5] See e.g. [64].

Appendix A
Classical and Mild Solutions of the Transformed RPDE

In this appendix, to make the expository as much self-contained as possible, a proof of Proposition 3.1 regarding the existence and uniqueness of a measurable global classical solution to the transformed Eq. (3.36) for any given $\mathcal{H}_\alpha$-valued (random) initial datum. We also introduce the definition of a mild solution to Eq. (3.36), and consider the existence and uniqueness of such mild solutions.

Proof of Proposition 3.1 We proceed in three steps. Since Eq. (3.36) becomes a non-autonomous PDE for each fixed ω, the existence of a unique solution for each given deterministic initial datum in $\mathcal{H}_\alpha$ with the announced regularity given in (3.37) can be proved by relying on the classical existence theory of solutions over finite time intervals and the half-line; see, e.g., [92, Thm. 3.3.3, Cor. 3.3.5]. The measurability property of such solutions requires more attention and details are provided in Step 2. The measurability property of solutions with random initial data as claimed right after Proposition 3.1 follows then from a basic composition argument as explained in Step 3.

Step 1. For each fixed ω, let us introduce

$$
\begin{aligned}
f_\omega(t, v) &:= z_\sigma(\theta_t\omega)v + G(\theta_t\omega, v) \\
&= z_\sigma(\theta_t\omega)v + e^{-z_\sigma(\theta_t\omega)} F(e^{z_\sigma(\theta_t\omega)}v), \quad \forall\, t \geq 0,\ v \in \mathcal{H}_\alpha.
\end{aligned}
$$

In order to apply the aforementioned non-autonomous theory, we first note that the following conditions hold naturally for $f_\omega(t, v)$:

$$
\begin{aligned}
&f_\omega \text{ is locally Hölder continuous in } t \\
&\quad \text{and locally Lipschitz in } v,\ \forall\, t \in \mathbb{R}^+,\ v \in \mathcal{H}_\alpha;
\end{aligned} \tag{A.1}
$$

and there exists a positive continuous function $Q_\omega\colon [0, \infty) \to \mathbb{R}^+$ depending on ω such that

$$
\|f_\omega(t, v)\| \leq Q_\omega(t)(1 + \|v\|_\alpha), \quad \forall\, t \geq 0,\ v \in \mathcal{H}_\alpha. \tag{A.2}
$$

M.D. Chekroun et al., *Approximation of Stochastic Invariant Manifolds*, SpringerBriefs in Mathematics, DOI 10.1007/978-3-319-12496-4

These are indeed a direct consequence of the fact that the nonlinearity $F\colon \mathscr{H}_\alpha \to \mathscr{H}$ is assumed to be globally Lipschitz, and that the function $t \mapsto z_\sigma(\theta_t\omega)$ is locally γ-Hölder continuous for all ω with any $\gamma \in (0, 1/2)$ according to Lemma 3.1.

The continuous dependence of the solutions with respect to the initial data and the parameter λ can be derived by using e.g. [92, Thm. 3.4.4].

Step 2. Now, we analyze the measurability of the solution for each fixed λ. First, we show that $v_\lambda(t, \cdot; v_0)$ is $(\mathscr{F}; \mathscr{B}(\mathscr{H}_\alpha))$-measurable for each fixed $t > 0$ and $v_0 \in \mathscr{H}_\alpha$. The treatment here is inspired by [27]. To this end, let us introduce for such given t and v_0 the space

$$X := C_{v_0}([0, t]; \mathscr{H}_\alpha) := \{v\colon [0, t] \to \mathscr{H}_\alpha \mid v \text{ is continuous and } v(0) = v_0\}. \tag{A.3}$$

Note that this space is a separable complete metric space when endowed with the metric

$$d_\gamma(v_1, v_2) := \sup_{t' \in [0,t]} e^{-\gamma t'} \|v_1(t') - v_2(t')\|_\alpha, \tag{A.4}$$

where $\gamma > 0$.

Now, for each fixed ω, let us introduce the following mapping $\mathscr{T}_{t,v_0}^{\omega,\lambda}$ defined on X:

$$\mathscr{T}_{t,v_0}^{\omega,\lambda}[v](s) := e^{sL_\lambda} v_0 + \int_0^s e^{(s-s')L_\lambda} f_\omega(s', v(s'))\mathrm{d}s', \quad \forall\, s \in [0, t],\ v \in X. \tag{A.5}$$

This mapping is well-defined due to (A.2) and the fact that there exists $a_\lambda > 0$ and $C_1 > 0$ such that

$$\|e^{tL_\lambda}\|_{L(\mathscr{H}, \mathscr{H}_\alpha)} \leq \frac{C_1}{t^\alpha} e^{a_\lambda t}, \qquad \forall\, t > 0; \tag{A.6}$$

see [143, Thm. 44.5] for a derivation of (A.6). The same reasons show that $\mathscr{T}_{t,v_0}^{\omega,\lambda}$ maps X into itself.

We show now that $\mathscr{T}_{t,v_0}^{\omega,\lambda}$ is a contraction mapping on the space (X, d_γ) for a sufficiently large γ depending on ω. First note that there exists $C_2 > 0$ such that the following inequality holds:

$$\|e^{tL_\lambda}\|_{L(\mathscr{H}_\alpha, \mathscr{H}_\alpha)} \leq C_2 e^{a_\lambda t}, \qquad \forall\, t \geq 0, \tag{A.7}$$

where a_λ is the same as given in (A.6); see again [143, Thm. 44.5].

By using (A.6), (A.7), and the fact that G and F have the same Lipschitz constant $\mathrm{Lip}(F)$, we obtain

$$
\begin{aligned}
&d_\gamma(\mathcal{T}^{\omega,\lambda}_{t,\nu_0}[\nu_1], \mathcal{T}^{\omega,\lambda}_{t,\nu_0}[\nu_2]) \\
&\quad \le \sup_{s\in[0,t]} e^{-\gamma s}\Bigg\{\int_0^s \|e^{(s-s')L_\lambda} z_\sigma(\theta_{s'}\omega)(\nu_1(s')-\nu_2(s'))\|_\alpha \mathrm{d}s' \\
&\qquad\qquad + \int_0^s \|e^{(s-s')L_\lambda}\big(G(\theta_{s'}\omega,\nu_1(s')) - G(\theta_{s'}\omega,\nu_2(s'))\big)\|_\alpha \mathrm{d}s'\Bigg\} \\
&\quad \le \sup_{s\in[0,t]} e^{-\gamma s}\Bigg(C_2\int_0^s e^{a_\lambda(s-s')}|z_\sigma(\theta_{s'}\omega)|\|\nu_1(s')-\nu_2(s')\|_\alpha \mathrm{d}s' \\
&\qquad\qquad + C_1 \mathrm{Lip}(F)\int_0^s \frac{e^{a_\lambda(s-s')}}{(s-s')^\alpha}\|\nu_1(s')-\nu_2(s')\|_\alpha \mathrm{d}s'\Bigg) \\
&\quad \le \sup_{s\in[0,t]} C d_\gamma(\nu_1,\nu_2)\int_0^s e^{(a_\lambda-\gamma)(s-s')}\left(|z_\sigma(\theta_{s'}\omega)| + \frac{\mathrm{Lip}(F)}{(s-s')^\alpha}\right)\mathrm{d}s',
\end{aligned}
\tag{A.8}
$$

where $C = \max\{C_1, C_2\}$.

We conclude about the contractive property of $\mathcal{T}^{\omega,\lambda}_{t,\nu_0}$ when γ is sufficiently large. First note that from the Lebesgue dominated convergence theorem the following property holds:

$$
\forall s \in [0,t],\ g_\gamma(s) := \int_0^s e^{(a_\lambda-\gamma)(s-s')}\left(|z_\sigma(\theta_{s'}\omega)| + \frac{\mathrm{Lip}(F)}{(s-s')^\alpha}\right)\mathrm{d}s' \underset{\gamma\to+\infty}{\longrightarrow} 0.
$$

Note also that for $\gamma' > \gamma > a_\lambda$, we have that $g_{\gamma'}(s) < g_\gamma(s)$ for all $s \in [0,t]$. Since $[0,t]$ is furthermore compact it follows from the Dini Theorem [68, Thm. 2.4.10] that g_γ converges uniformly to zero on $[0,t]$ as γ tends to $+\infty$. As a consequence, there exists $\gamma(\omega)$ sufficiently large, such that

$$
\begin{aligned}
\sup_{s\in[0,t]} C g_{\gamma(\omega)}(s) &= \sup_{s\in[0,t]} C\int_0^s e^{(a_\lambda-\gamma(\omega))(s-s')}\left(|z_\sigma(\theta_{s'}\omega)| + \frac{\mathrm{Lip}(F)}{(s-s')^\alpha}\right)\mathrm{d}s' \\
&< \frac{1}{2}.
\end{aligned}
\tag{A.9}
$$

For such a chosen $\gamma(\omega)$, we conclude then from (A.8)–(A.9) and the Banach fixed point theorem that there exists a unique fixed point $\nu^{\omega,\lambda}_{t,\nu_0} \in X$ of the operator $\mathcal{T}^{\omega,\lambda}_{t,\nu_0}$. Note also that the solution $\nu_\lambda(\cdot,\omega;\nu_0)$ obtained in Step 1, when restricted to the time interval $[0,t]$, is in X and is clearly a fixed point of $\mathcal{T}^{\omega,\lambda}_{t,\nu_0}$. By uniqueness of the fixed point, we have thus:

$$\nu_\lambda(s,\omega;\nu_0) = \nu^{\omega,\lambda}_{t,\nu_0}(s), \quad \forall\, s \in [0,t]. \tag{A.10}$$

Now we show that $\nu_\lambda(t,\cdot;\nu_0)$ is measurable. This results from the fact that $\nu_\lambda(t,\cdot;\nu_0)$ is obtained as the limit of a standard Picard scheme associated with $\mathscr{T}^{\cdot,\lambda}_{t,\nu_0}$ which maps $(\mathscr{B}([0,t])\otimes\mathscr{F};\mathscr{B}(\mathscr{H}_\alpha))$-measurable mappings to $(\mathscr{B}([0,t])\otimes\mathscr{F};\mathscr{B}(\mathscr{H}_\alpha))$-measurable mappings, where $\mathscr{B}([0,t])$ denotes the trace σ-algebra of $\mathscr{B}(\mathbb{R})$ restricted to $[0,t]$. For the sake of clarity, we provide the details of such a standard argument.

Let us define a constant mapping $\nu^0_{t,\nu_0}\colon [0,t]\times\Omega\to\mathscr{H}_\alpha$ by:

$$\nu^0_{t,\nu_0}(s,\omega) \equiv \nu_0, \quad \forall\, s\in[0,t],\ \omega\in\Omega. \tag{A.11}$$

Obviously, ν^0_{t,ν_0} is $(\mathscr{B}([0,t])\otimes\mathscr{F};\mathscr{B}(\mathscr{H}_\alpha))$-measurable and $\nu^0_{t,\nu_0}(\cdot,\omega)\in X$ for each ω. For each $n\geq 1$, let us define recursively $\nu^n_{t,\nu_0}\colon[0,t]\times\Omega\to\mathscr{H}_\alpha$ by the following Picard scheme:

$$\nu^n_{t,\nu_0}(s,\omega) := \mathscr{T}^{\omega,\lambda}_{t,\nu_0}[\nu^{n-1}_{t,\nu_0}(\cdot,\omega)](s), \quad \forall\, s\in[0,t],\ \omega\in\Omega. \tag{A.12}$$

As noted above, ν^n_{t,ν_0} is $(\mathscr{B}([0,t])\otimes\mathscr{F};\mathscr{B}(\mathscr{H}_\alpha))$-measurable for each n. It then follows that $\nu^n_{t,\nu_0}(t,\cdot)$ is $(\mathscr{F};\mathscr{B}(\mathscr{H}_\alpha))$-measurable for each $n\geq 0$.

Let $\gamma(\omega)>0$ be chosen such that $\mathscr{T}^{\omega,\lambda}_{t,\nu_0}$ is a contraction on $(X,d_{\gamma(\omega)})$. Then, we get from (A.10) that

$$d_{\gamma(\omega)}\big(\nu_\lambda(\cdot,\omega;\nu_0),\nu^n_{t,\nu_0}(\cdot,\omega)\big) = d_{\gamma(\omega)}\big(\nu^{\omega,\lambda}_{t,\nu_0}(\cdot),\nu^n_{t,\nu_0}(\cdot,\omega)\big)\to 0 \text{ as } n\to\infty,\ \forall\omega\in\Omega.$$

This together with the definition of the metric $d_{\gamma(\omega)}$ implies that the following pointwise limit exists

$$\nu_\lambda(t,\omega;\nu_0) = \lim_{n\to\infty}\nu^n_{t,\nu_0}(t,\omega), \quad \forall\,\omega\in\Omega,\ t>0, \tag{A.13}$$

where the limit is taken in $\mathscr{H}_\alpha$. Since $\mathscr{H}_\alpha$ is separable and the sequence $\nu^n_{t,\nu_0}(t,\cdot)$ is $(\mathscr{F};\mathscr{B}(\mathscr{H}_\alpha))$-measurable, we conclude from (A.13) that $\nu_\lambda(t,\cdot;\nu_0)\colon\Omega\to\mathscr{H}_\alpha$ is also measurable.

Since $\nu_\lambda(\cdot,\omega;\nu_0)$ is continuous for each ω; and $\nu_\lambda(t,\cdot;\nu_0)$ is measurable for each $t>0$, then $\nu_\lambda(\cdot,\cdot;\nu_0)$ is $\big(\mathscr{B}(\mathbb{R}^+)\otimes\mathscr{F};\mathscr{B}(\mathscr{H}_\alpha)\big)$-measurable by using for instance [32, LemmaIII.14].

Still based on [32, LemmaIII.14], since $\nu_\lambda(t,\omega;\cdot)$ is continuous for each ω and $t>0$, and $\nu_\lambda(\cdot,\cdot;\nu_0)$ is measurable for each ν_0, then ν_λ is $\big(\mathscr{B}(\mathbb{R}^+)\otimes\mathscr{F}\otimes\mathscr{B}(\mathscr{H}_\alpha);\mathscr{B}(\mathscr{H}_\alpha)\big)$-measurable.

Step 3. For any given measurable random initial datum $\nu_0(\omega)$, we can define ω-wisely the solution $\nu_{\lambda,\nu_0(\omega)}(t,\omega) := \nu_\lambda(t,\omega;\nu_0(\omega))$ as done in Step 1. Clearly, the solution has the regularity given in (3.37). Let

$$g\colon\mathbb{R}^+\times\Omega\to\mathbb{R}^+\times\Omega\times\mathscr{H}_\alpha, \quad (t,\omega)\mapsto(t,\omega,\nu_0(\omega)). \tag{A.14}$$

Then, we have $\nu_{\lambda,\nu_0(\omega)}(t,\omega) = \nu_\lambda \circ g(t,\omega)$. Since both g and ν_λ are measurable, then $\nu_{\lambda,\nu_0(\omega)}$ is also measurable. The proof is complete. □

Now, we introduce the following definition of mild solutions to Eq. (3.36).

Definition A.1 Let $\mathscr{H}_\alpha$ be the interpolation space associated with the fractional power A^α for some $\alpha \in (0,1)$, where A is given in (3.4). Let $J := [t_1, t_2]$ be a closed interval in $\mathbb{R}$, and λ be a fixed value in $\mathbb{R}$. A mapping $\nu_\lambda \colon J \times \Omega \to \mathscr{H}_\alpha$, which is jointly measurable and continuous in t, is said to be a mild solution of Eq. (3.36) in the space $\mathscr{H}_\alpha$ on J with initial datum $\nu_\lambda(t_1,\omega) = \nu_0$, if it satisfies the following integral equation:

$$\nu_\lambda(t,\omega) = \mathfrak{T}_{\lambda,\sigma}(t,t_1;\omega)\nu_0 + \int_{t_1}^{t} \mathfrak{T}_{\lambda,\sigma}(t,s;\omega)G(\theta_s\omega, \nu_\lambda(s,\omega))\,\mathrm{d}s, \quad \forall\, t \in J,\ \omega \in \Omega, \tag{A.15}$$

where $\mathfrak{T}_{\lambda,\sigma}$ is the solution operator associated with the linearized equation of Eq. (3.36) given in Sect. 3.4.

We call ν_λ a mild solution of Eq. (3.36) on $[t_1,\infty)$ if it is a mild solution on $[t_1,t_2]$ for any $t_2 > t_1$. Similarly, ν_λ is called a mild solution of Eq. (3.36) on $(-\infty, t_2]$ if it is a mild solution on $[t_1,t_2]$ for any $t_1 < t_2$.

In Appendix B, we will make use of the following elementary lemma.

Lemma A.1 *For any given $t_1 < t_2$, a measurable mapping $\nu_\lambda \colon [t_1,t_2]\times\Omega \to \mathscr{H}_\alpha$ is a mild solution to Eq. (3.36) if and only if it satisfies the following integral equation:*

$$\begin{aligned}\nu_\lambda(t,\omega) &= \mathfrak{T}_{\lambda,\sigma}(t,t_2;\omega)P_{\mathfrak{c}}\nu_\lambda(t_2,\omega) - \int_{t}^{t_2} \mathfrak{T}_{\lambda,\sigma}(t,s;\omega)P_{\mathfrak{c}}G(\theta_s\omega,\nu_\lambda(s,\omega))\,\mathrm{d}s\\ &\quad + \mathfrak{T}_{\lambda,\sigma}(t,t_1;\omega)P_{\mathfrak{s}}\nu_\lambda(t_1,\omega)\\ &\quad + \int_{t_1}^{t} \mathfrak{T}_{\lambda,\sigma}(t,s;\omega)P_{\mathfrak{s}}G(\theta_s\omega,\nu_\lambda(s,\omega))\,\mathrm{d}s, \quad t \in [t_1,t_2].\end{aligned} \tag{A.16}$$

Proof Assume that ν_λ is a mild solution of Eq. (3.36) on $[t_1,t_2]$, we check that it satisfies (A.16). First note that by the definition of mild solutions, we have

$$\nu_\lambda(t_2,\omega) = \mathfrak{T}_{\lambda,\sigma}(t_2,t_1;\omega)\nu_\lambda(t_1,\omega) + \int_{t_1}^{t_2} \mathfrak{T}_{\lambda,\sigma}(t_2,s;\omega)G(\theta_s\omega,\nu_\lambda(s,\omega))\,\mathrm{d}s, \tag{A.17}$$

which leads to

$$P_{\mathfrak{c}} v_\lambda(t_2, \omega) = \mathfrak{T}_{\lambda,\sigma}(t_2, t_1; \omega) P_{\mathfrak{c}} v_\lambda(t_1, \omega) + \int_{t_1}^{t_2} \mathfrak{T}_{\lambda,\sigma}(t_2, s; \omega) P_{\mathfrak{c}} G(\theta_s \omega, v_\lambda(s, \omega))\, ds, \tag{A.18}$$

where we used the fact that the solution operator $\mathfrak{T}_{\lambda,\sigma}$ leaves invariant the subspaces $\mathscr{H}^{\mathfrak{c}}$ and $\mathscr{H}^{\mathfrak{s}}$ as pointed out in Sect. 3.4.

We then obtain from the identity above that

$$\begin{aligned}
&\mathfrak{T}_{\lambda,\sigma}(t, t_2; \omega) P_{\mathfrak{c}} v_\lambda(t_2, \omega) \\
&= \mathfrak{T}_{\lambda,\sigma}(t, t_2; \omega) \mathfrak{T}_{\lambda,\sigma}(t_2, t_1; \omega) P_{\mathfrak{c}} v_\lambda(t_1, \omega) \\
&\quad + \mathfrak{T}_{\lambda,\sigma}(t, t_2; \omega) \int_{t_1}^{t_2} \mathfrak{T}_{\lambda,\sigma}(t_2, s; \omega) P_{\mathfrak{c}} G(\theta_s \omega, v_\lambda(s, \omega))\, ds \qquad \text{(A.19)} \\
&= \mathfrak{T}_{\lambda,\sigma}(t, t_1; \omega) P_{\mathfrak{c}} v_\lambda(t_1, \omega) \\
&\quad + \int_{t_1}^{t_2} \mathfrak{T}_{\lambda,\sigma}(t, s; \omega) P_{\mathfrak{c}} G(\theta_s \omega, v_\lambda(s, \omega))\, ds, \quad t \in [t_1, t_2].
\end{aligned}$$

By definition, we also have

$$v_\lambda(t, \omega) = \mathfrak{T}_{\lambda,\sigma}(t, t_1; \omega) v_\lambda(t_1, \omega) + \int_{t_1}^{t} \mathfrak{T}_{\lambda,\sigma}(t, s; \omega) G(\theta_s \omega, v_\lambda(s, \omega))\, ds, \quad t \in [t_1, t_2],$$

which can be rewritten as follows by simply using the fact that $P_{\mathfrak{c}} + P_{\mathfrak{s}} = \mathrm{Id}$:

$$\begin{aligned}
v_\lambda(t, \omega) &= \mathfrak{T}_{\lambda,\sigma}(t, t_1; \omega) P_{\mathfrak{c}} v_\lambda(t_1, \omega) \\
&\quad + \int_{t_1}^{t} \mathfrak{T}_{\lambda,\sigma}(t, s; \omega) P_{\mathfrak{c}} G(\theta_s \omega, v_\lambda(s, \omega))\, ds \\
&\quad + \mathfrak{T}_{\lambda,\sigma}(t, t_1; \omega) P_{\mathfrak{s}} v_\lambda(t_1, \omega) \qquad \text{(A.20)} \\
&\quad + \int_{t_1}^{t} \mathfrak{T}_{\lambda,\sigma}(t, s; \omega) P_{\mathfrak{s}} G(\theta_s \omega, v_\lambda(s, \omega))\, ds, \ t \in [t_1, t_2].
\end{aligned}$$

The expression of $\mathfrak{T}_{\lambda,\sigma}(t, t_1; \omega) P_{\mathfrak{c}} v_\lambda(t_1, \omega)$ provided implicitly by (A.19) leads then to:

$$\mathfrak{T}_{\lambda,\sigma}(t, t_1; \omega) P_c v_\lambda(t_1, \omega) + \int_{t_1}^{t} \mathfrak{T}_{\lambda,\sigma}(t, s; \omega) P_c G(\theta_s \omega, v_\lambda(s, \omega))\, ds$$

$$= \mathfrak{T}_{\lambda,\sigma}(t, t_2; \omega) P_c v_\lambda(t_2, \omega) - \int_{t_1}^{t_2} \mathfrak{T}_{\lambda,\sigma}(t, s; \omega) P_c G(\theta_s \omega, v_\lambda(s, \omega))\, ds$$

$$+ \int_{t_1}^{t} \mathfrak{T}_{\lambda,\sigma}(t, s; \omega) P_c G(\theta_s \omega, v_\lambda(s, \omega))\, ds$$

$$= \mathfrak{T}_{\lambda,\sigma}(t, t_2; \omega) P_c v_\lambda(t_2, \omega) - \int_{t}^{t_2} \mathfrak{T}_{\lambda,\sigma}(t, s; \omega) P_c G(\theta_s \omega, v_\lambda(s, \omega))\, ds.$$

Using this last identity in (A.20), we obtain then (A.16). The converse can be obtained in the same fashion, and we omit the details here. □

The existence and uniqueness problem of mild solutions to Eq. (3.36) is easily solved by relying on Proposition 3.1 and the following elementary result.

Proposition A.1 *Consider the RPDE* (3.36). *The assumptions on* L_λ *and* F *are those of Sect.* 3.1 *where* F *is assumed to be globally Lipschitz here; see* (3.7).

Then, for each $\lambda \in \mathbb{R}$ *and* $t_1 < t_2$, *a mapping* $v_\lambda \colon [t_1, t_2] \times \Omega \to \mathcal{H}_\alpha$ *is a mild solution of Eq.* (3.36) *on the time interval* $[t_1, t_2]$ *if and only if* v_λ *is a classical solution of Eq.* (3.36) *in the sense given in Proposition* 3.1.

Proof This is a direct generalization of [92, Lemma 3.3.2]. Indeed, by using the basic random change of variables $\widetilde{v}_\lambda(t, \omega) := e^{-\int_{t_1}^{t} z_\sigma(\theta_\tau \omega)\, d\tau \mathrm{Id}} v_\lambda(t, \omega)$, it can be checked that v_λ is a classical solution of Eq. (3.36) on (t_1, t_2) if and only if $\widetilde{v}_\lambda$ is a classical solution of the following equation

$$\frac{d\widetilde{v}}{dt} = L_\lambda \widetilde{v} + \widetilde{G}(\theta_t \omega, \widetilde{v}), \tag{A.21}$$

where $\widetilde{G}(\theta_t \omega, \widetilde{v}) = e^{-\int_{t_1}^{t} z_\sigma(\theta_\tau \omega)\, d\tau \mathrm{Id}} G(\theta_t \omega, e^{\int_{t_1}^{t} z_\sigma(\theta_\tau \omega)\, d\tau \mathrm{Id}} \widetilde{v})$. Note also that for each ω, the nonlinearity $\widetilde{G}(\theta_t \omega, \widetilde{v})$ clearly satisfies the conditions (A.1)–(A.2) with $\widetilde{G}$ in place of f_ω thanks to the Lipschitz property of G with respect to v and the Hölder continuity of the OU process $t \mapsto z_\sigma(\theta_t \omega)$. In particular, the conditions required in [92, Lemma 3.3.2] are met for Eq. (A.21), leading to the conclusion that $\widetilde{v}_\lambda$ is a classical solution to Eq. (A.21) on $[t_1, t_2]$ if and only if it is a mild solution, namely if and only if it satisfies the following integral equation:

$$\widetilde{v}_\lambda(t,\omega) = e^{(t-t_1)L_\lambda} v_0 + \int_{t_1}^{t} e^{(t-s)L_\lambda} \widetilde{G}(\theta_s\omega, \widetilde{v}_\lambda(s,\omega))\, \mathrm{d}s, \quad \forall\, t \in [t_1, t_2],\ \omega \in \Omega, \tag{A.22}$$

where $v_0 = \widetilde{v}_\lambda(t_1, \omega)$. By the construction of $\widetilde{v}_\lambda$, one readily sees that $\widetilde{v}_\lambda$ satisfies (A.22) if and only if v_λ satisfies (A.15). The proof is complete. □

Appendix B
Proof of Theorem 4.1

The proof is based on the Lyapunov-Perron method, and we adapt mainly the presentation of [66] to our functional setting. We split the proof into four steps. In Step 1, the sought random invariant manifold is characterized as the random set consisting of all elements in $\mathscr{H}_\alpha$ such that there exists a complete trajectory of Eq. (4.1) passing through each such element at $t = 0$, which has controlled growth as $t \to -\infty$. This characterization is shown to be equivalent to an integral equation to be satisfied by each such mild solution. The latter characterization via an integral equation is more suitable for analysis based on a fixed point argument relying on the partial-dichotomy estimates (3.46). A related fixed point problem associated with this integral equation and parameterized by the critical variable $\xi \in \mathscr{H}^{\mathfrak{c}}$ is analyzed in Step 2 by usage of the uniform contraction mapping principle [44, Theorems2.1–2.2]. In Step 3, by taking the projection onto the non-critical space $\mathscr{H}_\alpha^{\mathfrak{s}}$ of the solution (at $t = 0$) obtained in Step 2 for each given ξ, we build a random $\mathscr{H}_\alpha^{\mathfrak{s}}$-valued function defined on $\mathscr{H}^{\mathfrak{c}}$, whose graph gives the sought random invariant manifold; the invariance property of the manifold being examined in Step 4.

Proof of Theorem 4.1 We proceed in four steps as outlined above.

Step 1. Characterization of $\mathfrak{M}_\lambda(\omega)$ via an integral equation. For each $\lambda \in \Lambda$ and $\omega \in \Omega$, let $\mathfrak{M}_\lambda(\omega)$ be the subset of $\mathscr{H}_\alpha$ defined by

$$\begin{aligned} \mathfrak{M}_\lambda(\omega) := \big\{ u_0 \in \mathscr{H}_\alpha \mid \exists\, u_\lambda(\cdot, \omega; u_0) \in C_\eta^- \text{with}\, u_\lambda(0, \omega; u_0) = u_0, \\ \text{which is furthermore a mild solution of Eq. (4.1)} \\ \text{on } (-\infty, 0] \text{ in the sense of Definition A.1.} \big\}, \end{aligned} \tag{B.1}$$

where C_η^- is the space defined in (4.3) with η chosen according to condition (4.7).

Note that $\mathfrak{M}_\lambda(\omega)$ thus defined is not empty, since the origin is clearly contained in it. We will show in later steps that $\mathfrak{M}_\lambda$ so obtained is the global random invariant manifold that we look for. In the current step, we provide an equivalent characterization of this set. Given any $u_0 \in \mathfrak{M}_\lambda(\omega)$, let $u_\lambda(\cdot, \omega; u_0) \in C_\eta^-$ be a corresponding mild solution of Eq. (4.1) on $(-\infty, 0]$ with $u_\lambda(0, \omega; u_0) = u_0$. Namely, u_λ satisfies the following:

M.D. Chekroun et al., *Approximation of Stochastic Invariant Manifolds*, SpringerBriefs in Mathematics, DOI 10.1007/978-3-319-12496-4

$$u_\lambda(t,\omega;u_0)=\mathfrak{T}_{\lambda,\sigma}(t,\tau;\omega)u_\lambda(\tau,\omega;u_0)$$
$$+\int_\tau^t \mathfrak{T}_{\lambda,\sigma}(t,\tau;\omega)G(\theta_s\omega,u_\lambda(s,\omega;u_0))\,\mathrm{d}s,\quad\forall\ \tau\le t\le 0,\qquad\text{(B.2)}$$

where $\mathfrak{T}_{\lambda,\sigma}$ is the solution operator associated with the linearized stochastic flow as described in Sect. 3.4.

According to Lemma A.1, Eq. (B.2) can be rewritten as

$$u_\lambda(t,\omega;u_0)=\mathfrak{T}_{\lambda,\sigma}(t,0;\omega)P_{\mathfrak{c}}u_0-\int_t^0 \mathfrak{T}_{\lambda,\sigma}(t,s;\omega)P_{\mathfrak{c}}G(\theta_s\omega,u_\lambda(s,\omega;u_0))\,\mathrm{d}s$$
$$+\mathfrak{T}_{\lambda,\sigma}(t,\tau;\omega)P_{\mathfrak{s}}u_\lambda(\tau,\omega;u_0)\qquad\text{(B.3)}$$
$$+\int_\tau^t \mathfrak{T}_{\lambda,\sigma}(t,s;\omega)P_{\mathfrak{s}}G(\theta_s\omega,u_\lambda(s,\omega;u_0))\,\mathrm{d}s,\quad\forall\ \tau\le t\le 0.$$

Projecting Eq. (B.3) onto the subspace $\mathscr{H}^{\mathfrak{c}}$, we obtain that

$$P_{\mathfrak{c}}u_\lambda(t,\omega;u_0)=\mathfrak{T}_{\lambda,\sigma}(t,0;\omega)P_{\mathfrak{c}}u_0$$
$$-\int_t^0 \mathfrak{T}_{\lambda,\sigma}(t,s;\omega)P_{\mathfrak{c}}G(\theta_s\omega,u_\lambda(s,\omega;u_0))\,\mathrm{d}s,\qquad\text{(B.4)}$$

where we used the fact that $P_{\mathfrak{c}}$ commutes with L_λ and hence with $\mathfrak{T}_{\lambda,\sigma}$.

Projecting Eq. (B.3) onto the subspace $\mathscr{H}^{\mathfrak{s}}_\alpha$, we obtain that

$$P_{\mathfrak{s}}u_\lambda(t,\omega;u_0)=\mathfrak{T}_{\lambda,\sigma}(t,\tau;\omega)P_{\mathfrak{s}}u_\lambda(\tau,\omega;u_0)$$
$$+\int_\tau^t \mathfrak{T}_{\lambda,\sigma}(t,s;\omega)P_{\mathfrak{s}}G(\theta_s\omega,u_\lambda(s,\omega;u_0))\,\mathrm{d}s.\qquad\text{(B.5)}$$

Since $u_\lambda(\cdot,\omega;u_0)\in C_\eta^-$, using (3.46a) we obtain for any $\tau\le t\le 0$ that

$$\|\mathfrak{T}_{\lambda,\sigma}(t,\tau;\omega)P_{\mathfrak{s}}u_\lambda(\tau,\omega;u_0)\|_\alpha$$
$$\le Ke^{\eta_2(t-\tau)+\int_\tau^t z_\sigma(\theta_s\omega)\,\mathrm{d}s}\|u_\lambda(\tau,\omega;u_0)\|_\alpha\qquad\text{(B.6)}$$
$$\le Ke^{\eta_2 t-\int_t^0 z_\sigma(\theta_s\omega)\,\mathrm{d}s}e^{(\eta-\eta_2)\tau}\|u_\lambda(\cdot,\omega;u_0)\|_{C_\eta^-}.$$

By recalling that $\eta_2 < \eta$ from condition (4.7), we get that, for each $t \le 0$, the RHS above converges to zero as τ goes to $-\infty$. Now, by taking the limit $\tau \to -\infty$ in (B.5), we obtain that

$$P_{\mathfrak{s}} u_\lambda(t, \omega; u_0) = \int_{-\infty}^{t} \mathfrak{T}_{\lambda,\sigma}(t, s; \omega) P_{\mathfrak{s}} G\big(\theta_s \omega, u_\lambda(s, \omega; u_0)\big)\, ds, \quad \forall\, t \le 0. \tag{B.7}$$

For each $\xi \in \mathscr{H}^{\mathfrak{c}}$, $\lambda \in \Lambda$, and $\omega \in \Omega$, we define an operator $\mathscr{N}_\xi^{\omega,\lambda}$ as follows:

$$\begin{aligned} \mathscr{N}_\xi^{\omega,\lambda}[u](t) := {} & \mathfrak{T}_{\lambda,\sigma}(t, 0; \omega)\xi \\ & - \int_{t}^{0} \mathfrak{T}_{\lambda,\sigma}(t, s; \omega) P_{\mathfrak{c}} G(\theta_s \omega, u(s))\, ds \\ & + \int_{-\infty}^{t} \mathfrak{T}_{\lambda,\sigma}(t, s; \omega) P_{\mathfrak{s}} G(\theta_s \omega, u(s))\, ds, \quad t \le 0,\ u \in C_\eta^-. \end{aligned} \tag{B.8}$$

Combining (B.4) and (B.7), we obtain that u_λ is a fixed point of $\mathscr{N}_\xi^{\omega,\lambda}$ with $\xi = P_{\mathfrak{c}} u_0$. Conversely, if $u_\lambda(\cdot, \omega) \in C_\eta^-$ is a fixed point of (B.8) with $u_\lambda(0, \omega) = u_0$, we show in the following that it is also a mild solution to Eq. (4.1) on $(-\infty, 0]$.

According to Lemma A.1, we only need to show that for any $\tau < 0$, $u_\lambda(t, \omega)$ can be written into the form given in (B.3) for any $t \in [\tau, 0]$.

First note that by applying $P_{\mathfrak{s}}$ to both sides of (B.8) and setting $t = \tau$, we obtain that

$$P_{\mathfrak{s}} u_\lambda(\tau, \omega) = P_{\mathfrak{s}} \mathscr{N}_\xi^{\omega,\lambda}[u_\lambda](\tau) = \int_{-\infty}^{\tau} \mathfrak{T}_{\lambda,\sigma}(\tau, s; \omega) P_{\mathfrak{s}} G(\theta_s \omega, u_\lambda(s, \omega))\, ds. \tag{B.9}$$

Note also that for any $t \in [\tau, 0]$, we have

$$\begin{aligned} & \int_{-\infty}^{t} \mathfrak{T}_{\lambda,\sigma}(t, s; \omega) P_{\mathfrak{s}} G(\theta_s \omega, u_\lambda(s, \omega))\, ds \\ & \qquad = \int_{\tau}^{t} \mathfrak{T}_{\lambda,\sigma}(t, s; \omega) P_{\mathfrak{s}} G(\theta_s \omega, u_\lambda(s, \omega))\, ds \\ & \qquad\quad + \int_{-\infty}^{\tau} \mathfrak{T}_{\lambda,\sigma}(t, s; \omega) P_{\mathfrak{s}} G(\theta_s \omega, u_\lambda(s, \omega))\, ds, \end{aligned}$$

and

$$\int_{-\infty}^{\tau} \mathfrak{T}_{\lambda,\sigma}(t,s;\omega) P_{\mathfrak{s}} G(\theta_s\omega, u_\lambda(s,\omega))\,\mathrm{d}s$$
$$= \mathfrak{T}_{\lambda,\sigma}(t,\tau;\omega) \int_{-\infty}^{\tau} \mathfrak{T}_{\lambda,\sigma}(\tau,s;\omega) P_{\mathfrak{s}} G(\theta_s\omega, u_\lambda(s,\omega))\,\mathrm{d}s$$
$$= \mathfrak{T}_{\lambda,\sigma}(t,\tau;\omega) P_{\mathfrak{s}} u_\lambda(\tau,\omega),$$

where we used (B.9) to derive the last equality above.

By combining the above two identities, we obtain

$$\int_{-\infty}^{t} \mathfrak{T}_{\lambda,\sigma}(t,s;\omega) P_{\mathfrak{s}} G(\theta_s\omega, u_\lambda(s,\omega))\,\mathrm{d}s = \mathfrak{T}_{\lambda,\sigma}(t,\tau;\omega) P_{\mathfrak{s}} u_\lambda(\tau,\omega) + \int_{\tau}^{t} \mathfrak{T}_{\lambda,\sigma}(t,s;\omega) P_{\mathfrak{s}} G(\theta_s\omega, u_\lambda(s,\omega))\,\mathrm{d}s.$$

Noting that $\xi = P_{\mathfrak{c}} u_0$, (B.3) follows then by using the above identity in the following fixed point equation satisfied by u_λ:

$$\begin{aligned} u_\lambda(t,\omega) &= \mathcal{N}_\xi^{\omega,\lambda}[u_\lambda](t) \\ &= \mathfrak{T}_{\lambda,\sigma}(t,s;\omega)\xi - \int_t^0 \mathfrak{T}_{\lambda,\sigma}(t,s;\omega) P_{\mathfrak{c}} G(\theta_s\omega, u_\lambda(s,\omega))\,\mathrm{d}s \\ &\quad + \int_{-\infty}^{t} \mathfrak{T}_{\lambda,\sigma}(t,s;\omega) P_{\mathfrak{s}} G(\theta_s\omega, u_\lambda(s,\omega))\,\mathrm{d}s. \end{aligned} \tag{B.10}$$

Hence, u_λ is indeed a mild solution to Eq. (4.1) on $(-\infty, 0]$ thanks to Lemma A.1.

Consequently, we obtain the following equivalent characterization of the set $\mathfrak{M}_\lambda(\omega)$ defined in (B.1), that is for each $\omega \in \Omega$ and $\lambda \in \Lambda$,

$$\Big(u_0 \in \mathfrak{M}_\lambda(\omega)\Big) \Longleftrightarrow \Big(\exists\, u_\lambda(\cdot,\omega) \in C_\eta^- \text{ s.t. } u_\lambda(0,\omega) = u_0,\ u_\lambda = \mathcal{N}_{P_{\mathfrak{c}} u_0}^{\omega,\lambda}[u_\lambda]\Big), \tag{B.11}$$

where $\mathcal{N}_{P_{\mathfrak{c}} u_0}^{\omega,\lambda}$ is defined in (B.8) with $\xi = P_{\mathfrak{c}} u_0$.

Step 2. Unique fixed point of $\mathcal{N}_\xi^{\omega,\lambda}$. We show in this step that for each $\xi \in \mathcal{H}^{\mathfrak{c}}$, $\lambda \in \Lambda$, and $\omega \in \Omega$ the operator $\mathcal{N}_\xi^{\omega,\lambda}$ defined in (B.8) has a unique fixed point in the space C_η^-.

We first check that

$$\mathscr{N}_{\xi}^{\omega,\lambda} C_{\eta}^{-} \subset C_{\eta}^{-}. \tag{B.12}$$

Note that by the partial-dichotomy estimate in (3.46c) and the assumption that $\eta \in (\eta_2, \eta_1)$, we obtain that

$$\begin{aligned} &\sup_{t\in(-\infty,0]} e^{-\eta t + \int_t^0 z_\sigma(\theta_s\omega)\mathrm{d}s} \left\| \mathfrak{T}_{\lambda,\sigma}(t,0;\omega)\xi \right\|_\alpha \\ &\quad \le K \sup_{t\in(-\infty,0]} e^{-\eta t} e^{\eta_1 t} \|\xi\|_\alpha \\ &\quad \le K\|\xi\|_\alpha, \quad \forall\, \xi \in \mathscr{H}^{\mathfrak{c}}; \end{aligned}$$

and that

$$\begin{aligned} &\sup_{t\in(-\infty,0]} e^{-\eta t + \int_t^0 z_\sigma(\theta_s\omega)\mathrm{d}s} \left\| \int_t^0 \mathfrak{T}_{\lambda,\sigma}(t,s;\omega) P_{\mathfrak{c}} G(\theta_s\omega, u(s))\,\mathrm{d}s \right\|_\alpha \\ &\le \sup_{t\in(-\infty,0]} K\mathrm{Lip}(F)\|u\|_{C_\eta^-} \int_t^0 e^{(\eta_1-\eta)(t-s)}\,\mathrm{d}s \\ &= \frac{K\mathrm{Lip}(F)}{\eta_1-\eta}\|u\|_{C_\eta^-}, \qquad \forall\, u \in C_\eta^-. \end{aligned}$$

Similarly, by (3.46b), we obtain that

$$\begin{aligned} &\sup_{t\in(-\infty,0]} e^{-\eta t + \int_t^0 z_\sigma(\theta_s\omega)\mathrm{d}s} \left\| \int_{-\infty}^t \mathfrak{T}_{\lambda,\sigma}(t,s;\omega) P_{\mathfrak{s}} G(\theta_s\omega, u(s))\,\mathrm{d}s \right\|_\alpha \\ &\qquad \le K\mathrm{Lip}(F)\Gamma(1-\alpha)(\eta-\eta_2)^{\alpha-1}\|u\|_{C_\eta^-}. \end{aligned}$$

Using the above three estimates and the definition of $\mathscr{N}_{\xi}^{\omega,\lambda}$, we get

$$\begin{aligned} \|\mathscr{N}_{\xi}^{\omega,\lambda}[u]\|_{C_\eta^-} &= \sup_{t\in(-\infty,0]} e^{-\eta t + \int_t^0 z_\sigma(\theta_s\omega)\mathrm{d}s} \|\mathscr{N}_{\xi}^{\omega,\lambda}[u](t)\|_\alpha \\ &\le K\|\xi\|_\alpha + \Upsilon_1(F)\|u\|_{C_\eta^-}, \end{aligned} \tag{B.13}$$

where $\Upsilon_1(F)$ is defined in (4.7). Thus, (B.12) follows.

Note that for any $u_1, u_2 \in C_\eta^-$ we have that

$$\begin{aligned}
&\|\mathscr{N}_\xi^{\omega,\lambda}[u_1] - \mathscr{N}_\xi^{\omega,\lambda}[u_2]\|_{C_\eta^-} \\
&\leq \sup_{t\in(-\infty,0]} \Bigg\{ e^{-\eta t + \int_t^0 z_\sigma(\theta_s\omega)\,\mathrm{d}s} \Bigg(\Big\| \int_t^0 \mathfrak{T}_{\lambda,\sigma}(t,s;\omega) P_{\mathfrak{c}}\big(G(\theta_s\omega,u_1) - G(\theta_s\omega,u_2)\big)\,\mathrm{d}s \Big\|_\alpha \\
&\qquad + \Big\| \int_{-\infty}^t \mathfrak{T}_{\lambda,\sigma}(t,s;\omega) P_{\mathfrak{s}}\big(G(\theta_s\omega,u_1) - G(\theta_s\omega,u_2)\big)\,\mathrm{d}s \Big\|_\alpha \Bigg) \Bigg\} \\
&\leq \sup_{t\in(-\infty,0]} \Bigg\{ K\mathrm{Lip}(F)\|u_1 - u_2\|_{C_\eta^-} \Bigg(\int_t^0 e^{(\eta_1-\eta)(t-s)}\,\mathrm{d}s + \int_{-\infty}^t \frac{e^{(\eta_2-\eta)(t-s)}}{(t-s)^\alpha}\,\mathrm{d}s \Bigg) \Bigg\} \\
&\leq K\mathrm{Lip}(F)\big((\eta_1-\eta)^{-1} + \Gamma(1-\alpha)(\eta-\eta_2)^{\alpha-1}\big)\|u_1 - u_2\|_{C_\eta^-} \\
&= \Upsilon_1(F)\|u_1 - u_2\|_{C_\eta^-}.
\end{aligned} \tag{B.14}$$

This together with assumption (4.7) implies that $\mathscr{N}_\xi^{\omega,\lambda}$ is a contraction mapping on C_η^- with constant of contraction that is independent of ω, $\lambda \in \Lambda$ and $\xi \in \mathscr{H}^{\mathfrak{c}}$. Note also that by definition, $\mathscr{N}_\xi^{\omega,\lambda}$ is clearly Lipschitz in ξ, and for each fixed $u \in C_\eta^-$ the following estimate holds:

$$\|\mathscr{N}_{\xi_1}^{\omega,\lambda}[u] - \mathscr{N}_{\xi_2}^{\omega,\lambda}[u]\|_{C_\eta^-} \leq K\|\xi_1 - \xi_2\|_\alpha, \quad \forall\, \xi_1, \xi_2 \in \mathscr{H}^{\mathfrak{c}}. \tag{B.15}$$

We can apply now the uniform contraction mapping principle (see, e.g., [44, Theorems2.1–2.2]), which ensures that for each $\xi \in \mathscr{H}^{\mathfrak{c}}$, the mapping $\mathscr{N}_\xi^{\omega,\lambda}$ has a unique fixed point $u_{\lambda,\xi}(\cdot,\omega) \in C_\eta^-$. Moreover, it follows from (B.14) and (B.15) that the mapping $\xi \to u_{\lambda,\xi}(\cdot,\omega)$ is Lipschitz from $\mathscr{H}^{\mathfrak{c}}$ to C_η^-:

$$\begin{aligned}
&\|u_{\lambda,\xi_1}(\cdot,\omega) - u_{\lambda,\xi_2}(\cdot,\omega)\|_{C_\eta^-} \\
&\quad = \|\mathscr{N}_{\xi_1}^{\omega,\lambda}[u_{\lambda,\xi_1}] - \mathscr{N}_{\xi_2}^{\omega,\lambda}[u_{\lambda,\xi_2}]\|_{C_\eta^-} \\
&\quad \leq \frac{K\|\xi_1 - \xi_2\|_\alpha}{1 - \Upsilon_1(F)}.
\end{aligned} \tag{B.16}$$

Now we show that the mapping $(t,\omega,\xi) \mapsto u_{\lambda,\xi}(t,\omega)$ is jointly measurable for each fixed $\lambda \in \Lambda$. The argument is similar to the one used in Step 2 of the proof of Proposition 3.1 and relies here on the Picard scheme associated with $u = \mathscr{N}_{P_{\mathfrak{c}}u_0}^{\omega,\lambda}[u]$ where $u_0 = u(0,\omega)$. Let us denote the zero mapping from $(-\infty,0] \times \Omega$ to $\mathscr{H}_\alpha$ by Z; and define $u_n \colon (-\infty,0] \times \Omega \to \mathscr{H}_\alpha$ for each $n \geq 1$ to be:

$$u_n(t,\omega) := (\mathscr{N}_\xi^{\omega,\lambda})^n[Z(\cdot,\omega)](t), \quad t \leq 0,\ \omega \in \Omega. \tag{B.17}$$

Since $u_{\lambda,\xi}(\cdot, \omega)$ is obtained as the fixed point of $\mathcal{N}_\xi^{\omega,\lambda}$ in C_η^- for each fixed ω, and $Z(\cdot, \omega)$ is obviously in C_η^-, we infer that

$$u_{\lambda,\xi}(\cdot, \omega) = \lim_{n\to\infty} u_n(\cdot, \omega), \quad \forall\, \omega \in \Omega, \tag{B.18}$$

where the limit is taken in C_η^-.

Note furthermore that the operator $\mathcal{N}_\xi^{\cdot,\lambda}$ maps $(\mathscr{B}((-\infty, 0]) \otimes \mathscr{F}; \mathscr{B}(\mathscr{H}_\alpha))$-measurable mappings to $(\mathscr{B}((-\infty, 0]) \otimes \mathscr{F}; \mathscr{B}(\mathscr{H}_\alpha))$-measurable mappings; where $\mathscr{B}((-\infty, 0])$ denotes the trace σ-algebra of $\mathscr{B}(\mathbb{R})$ with respect to $(-\infty, 0]$. So, each u_n has this measurability. As a consequence, $u_{\lambda,\xi}$ is also $(\mathscr{B}((-\infty, 0]) \otimes \mathscr{F}; \mathscr{B}(\mathscr{H}_\alpha))$-measurable according to (B.18). Moreover, since $u_{\lambda,\xi}(t, \omega)$ is Lipschitz in ξ (hence continuous in ξ), by using for instance [32, LemmaIII.14], we conclude that $u_{\lambda,\xi}(t, \omega)$ is jointly measurable in t, ξ and ω for each fixed $\lambda \in \Lambda$.

Step 3. Construction of the random invariant manifold $\mathfrak{M}_\lambda$. Now, we show that the set $\mathfrak{M}_\lambda$ defined in (B.1) is indeed a random invariant manifold as the graph of a random $\mathscr{H}_\alpha^{\mathfrak{s}}$-valued function.

Let

$$h_\lambda(\xi, \omega) := P_{\mathfrak{s}} u_{\lambda,\xi}(0, \omega), \quad \forall\, \xi \in \mathscr{H}^{\mathfrak{c}},\ \omega \in \Omega. \tag{B.19}$$

Note that $h_\lambda(\xi, \omega)$ lives in $\mathscr{H}_\alpha^{\mathfrak{s}}$ because $u_{\lambda,\xi}(0, \omega) \in \mathscr{H}_\alpha$ by construction.

By the measurability property of $u_{\lambda,\xi}$ derived in the previous step, h_λ is measurable. Since $u_{\lambda,\xi}(\cdot, \omega)$ is the fixed point of $\mathcal{N}_\xi^{\omega,\lambda}$, we have in particular that

$$u_{\lambda,\xi}(0, \omega) = \mathcal{N}_\xi^{\omega,\lambda}[u_{\lambda,\xi}(\cdot, \omega)](0), \quad \forall\, \omega \in \Omega. \tag{B.20}$$

Therefore, by applying $P_{\mathfrak{s}}$ to (B.8) and setting t to zero, we get from (B.19):

$$h_\lambda(\xi, \omega) = \int_{-\infty}^{0} \mathfrak{T}_{\lambda,\sigma}(0, s; \omega) P_{\mathfrak{s}} G(\theta_s\omega, u_{\lambda,\xi}(s, \omega))\, \mathrm{d}s, \tag{B.21}$$

where we used the fact that the solution operator $\mathfrak{T}_{\lambda,\sigma}$ leaves invariant the subspaces $\mathscr{H}^{\mathfrak{c}}$ and $\mathscr{H}^{\mathfrak{s}}$ as pointed out in Sect. 3.4. By applying $P_{\mathfrak{c}}$ to (B.8) we get also:

$$\xi = P_{\mathfrak{c}} u_{\lambda,\xi}(0, \omega). \tag{B.22}$$

Now, (B.19) and (B.22) lead to

$$u_{\lambda,\xi}(0, \omega) = \xi + h_\lambda(\xi, \omega). \tag{B.23}$$

Recalling that $u_{\lambda,\xi}(\cdot, \omega) \in C_\eta^-$, we conclude from (B.23) and the equivalent characterization of $\mathfrak{M}_\lambda(\omega)$ given in (B.11) that

$$\mathfrak{M}_\lambda(\omega) = \{\xi + h_\lambda(\xi, \omega) \mid \xi \in \mathscr{H}^{\mathfrak{c}}\}. \tag{B.24}$$

Namely, $\mathfrak{M}_\lambda(\omega)$ is the graph of $h_\lambda(\cdot, \omega)$ for all ω.

We derive now the properties that h_λ satisfies as stated in the theorem. First note that $u_{\lambda,\xi}(t, \omega) \equiv 0$ if $\xi = 0$. This together with (B.21) and the fact that $G(\omega, 0) = 0$ for all ω implies that $h_\lambda(0, \omega) = 0$. By construction, h_λ depends continuously on λ for $\lambda \in \Lambda$ since $u_{\lambda,\xi}$ does. From (B.21), we get for any $\xi_1, \xi_2 \in \mathscr{H}^{\mathfrak{c}}$ that

$$\begin{aligned} &h_\lambda(\xi_1, \omega) - h_\lambda(\xi_2, \omega) \\ &= \int_{-\infty}^{0} \mathfrak{T}_{\lambda,\sigma}(0, s; \omega) P_{\mathfrak{s}}\Big(G(\theta_s\omega, u_{\lambda,\xi_1}(s, \omega)) - G(\theta_s\omega, u_{\lambda,\xi_2}(s, \omega))\Big) \mathrm{d}s. \end{aligned}$$

It then follows from (B.16) and (3.46b) that $h_\lambda(\cdot, \omega)$ is Lipschitz in ξ for each ω:

$$\begin{aligned} &\|h_\lambda(\xi_1, \omega) - h_\lambda(\xi_2, \omega)\|_\alpha \\ &\le \frac{K^2 \mathrm{Lip}(F)(\eta - \eta_2)^{\alpha-1}\Gamma(1-\alpha)}{1 - \Upsilon_1(F)} \|\xi_1 - \xi_2\|_\alpha, \quad \forall\, \xi_1, \xi_2 \in \mathscr{H}^{\mathfrak{c}}. \end{aligned} \tag{B.25}$$

The estimate of the Lipschitz constant in (4.9) follows now from (B.25).

By recalling from Step 1 that the fixed point $u_{\lambda,\xi}(\cdot, \omega)$ of $\mathscr{N}_\xi^{\omega,\lambda}$ is also the mild solution in C_η^- to Eq. (4.1) on $(-\infty, 0]$ with $u_{\lambda,\xi}(0, \omega) = \xi + h_\lambda(\xi, \omega)$, the integral representation (4.8) follows then from (B.21) and the definition of $\mathfrak{T}_{\lambda,\sigma}$ provided in Sect. 3.4.

Now we show that $\mathfrak{M}_\lambda$ is a random closed set. Since h_λ is continuous in ξ, according to (B.24), $\mathfrak{M}_\lambda(\omega)$ is a closed subset of $\mathscr{H}_\alpha$ for each ω. Then, by using the Selection Theorem [32, Thm. III.9] or [54, Thm. 2.6], in order to show that $\mathfrak{M}_\lambda$ is a random closed set, we only need to show that there exists a sequence $\{\gamma_n\}_{n\in\mathbb{N}}$ of measurable mappings $\gamma_n : \Omega \to \mathscr{H}_\alpha$, such that

$$\mathfrak{M}_\lambda(\omega) = \overline{\{\gamma_n(\omega) \| n \in \mathbb{N}\}}^{\mathscr{H}_\alpha}, \quad \forall\, \omega \in \Omega.$$

Since $\mathscr{H}_\alpha$ is separable, there exists a sequence $\{u_n\} \in (\mathscr{H}_\alpha)^{\mathbb{N}}$ which is dense in $\mathscr{H}_\alpha$. Now, for each u_n, let us define $\gamma_n : \Omega \to \mathscr{H}_\alpha$ as follows

$$\omega \mapsto \gamma_n(\omega) := P_{\mathfrak{c}} u_n + h_\lambda(P_{\mathfrak{c}} u_n, \omega).$$

Clearly, γ_n is measurable because $h_\lambda(P_{\mathfrak{c}} u, \cdot)$ is measurable for any fixed $u \in \mathscr{H}_\alpha$.

Since $\mathfrak{M}_\lambda(\omega)$ is closed for each $\omega \in \Omega$, we have by the construction of γ_n that

$$\overline{\{\gamma_n(\omega) \| n \in \mathbb{N}\}}^{\mathscr{H}_\alpha} \subset \mathfrak{M}_\lambda(\omega), \quad \forall\, \omega \in \Omega.$$

Now, we show that the reverse inclusion holds. Let ω be fixed in Ω. For each $u \in \mathfrak{M}_\lambda(\omega)$, it can be written as $u = P_{\mathfrak{c}}u + h_\lambda(P_{\mathfrak{c}}u, \omega)$. By the definition of $\{u_n\}$, there exists a subsequence, stilled denoted by $\{u_n\}$, which converges to u, leading to the convergence of $\{P_{\mathfrak{c}}u_n\}$ to $P_{\mathfrak{c}}u$. Then, by the continuity of $h_\lambda(\cdot, \omega)$ and the definition of γ_n, we have that $\{\gamma_n(\omega)\}$ converges to u. It follows that

$$\mathfrak{M}_\lambda(\omega) \subset \overline{\{\gamma_n\omega \mid n \in \mathbb{N}^*\}}^{\mathcal{H}_\alpha}, \quad \forall\, \omega \in \Omega.$$

We have thus proved that $\mathfrak{M}_\lambda$ is a random closed set.

Step 4. Invariance property of $\mathfrak{M}_\lambda$. We show in this last step that $\mathfrak{M}_\lambda$ is invariant, i.e.,

$$S_\lambda(t, \omega)\mathfrak{M}_\lambda(\omega) \subset \mathfrak{M}_\lambda(\theta_t\omega), \quad \forall\, t > 0,\ \omega \in \Omega, \tag{B.26}$$

where S_λ is the RDS associated with Eq. (4.1).

For each fixed ω, $u_0 \in \mathfrak{M}_\lambda(\omega)$, and $t > 0$, let $u_\lambda(s, \omega; u_0)$, $s \in [0, t]$, be the mild solution to Eq. (4.1) with initial datum u_0 (in the fiber ω). By Proposition A.1, we know that this mild solution exists, and is also a classical solution. Thus, according to the definition of S_λ, we have

$$S_\lambda(t, \omega)u_0 = u_\lambda(t, \omega; u_0). \tag{B.27}$$

Our goal is then to show that

$$S_\lambda(t, \omega)u_0 \in \mathfrak{M}_\lambda(\theta_t\omega). \tag{B.28}$$

Since $u_0 \in \mathfrak{M}_\lambda(\omega)$, from the characterization of $\mathfrak{M}_\lambda(\omega)$ provided in (B.1), we can extend the above mild solution to the interval $(-\infty, t]$ and the following property holds:

$$u_\lambda(s, \omega; u_0)|_{s\in(-\infty,0]} \in C_\eta^-(\omega). \tag{B.29}$$

Now, let

$$v_\lambda(s, \theta_t\omega; u_\lambda(t, \omega; u_0)) := u_\lambda(s + t, \omega; u_0), \quad \forall\, s \le 0. \tag{B.30}$$

Note that $v_\lambda(s, \theta_t\omega; u_\lambda(t, \omega; u_0))$ is a mild solution to Eq. (4.1) on $(-\infty, 0]$ which takes value $u_\lambda(t, \omega; u_0)$ in the fiber $\theta_t\omega$ when $s = 0$. Note also that

$$v_\lambda(0, \theta_t\omega; u_\lambda(t, \omega; u_0)) = u_\lambda(t, \omega; u_0) = S_\lambda(t, \omega)u_0. \tag{B.31}$$

Then, according to the characterization provided in (B.1) adapted to $\mathfrak{M}_\lambda(\theta_t\omega)$, in order to check (B.28), we only need to show that

$$v_\lambda(\cdot, \theta_t\omega; u_\lambda(t, \omega; u_0)) \in C_\eta^-(\theta_t\omega). \tag{B.32}$$

Note that

$$
\begin{aligned}
&\|v_\lambda(\cdot,\theta_t\omega;u_\lambda(t,\omega;u_0))\|_{C_\eta^-(\theta_t\omega)} \\
&\qquad = \sup_{s\le 0} e^{-\eta s+\int_s^0 z_\sigma(\theta_{\tau+t}\omega)\mathrm{d}\tau}\|v_\lambda(s,\theta_t\omega;u_\lambda(t,\omega;u_0))\|_\alpha \\
&\qquad = \sup_{s\le 0} e^{-\eta s+\int_{s+t}^t z_\sigma(\theta_\tau\omega)\mathrm{d}\tau}\|u_\lambda(s+t,\omega;u_0)\|_\alpha \\
&\qquad = \sup_{s'\le t} e^{-\eta(s'-t)+\int_{s'}^t z_\sigma(\theta_\tau\omega)\mathrm{d}\tau}\|u_\lambda(s',\omega;u_0)\|_\alpha \\
&\qquad = e^{\eta t+\int_0^t z_\sigma(\theta_\tau\omega)\mathrm{d}\tau}\sup_{s'\le t} e^{-\eta s'+\int_{s'}^0 z_\sigma(\theta_\tau\omega)\mathrm{d}\tau}\|u_\lambda(s',\omega;u_0)\|_\alpha .
\end{aligned}
\tag{B.33}
$$

Since $u_\lambda(s,\omega;u_0)$ is continuous on $[0,t]$, it clearly holds that

$$
\sup_{s'\in[0,t]} e^{-\eta s'+\int_{s'}^0 z_\sigma(\theta_\tau\omega)\mathrm{d}\tau}\|u_\lambda(s',\omega;u_0)\|_\alpha < \infty. \tag{B.34}
$$

The control on $(-\infty,0]$ by some finite constant is achieved thanks to (B.29). From (B.33), we obtain then that $\|v_\lambda(\cdot,\theta_t\omega;u_\lambda(t,\omega;u_0))\|_{C_\eta^-(\theta_t\omega)} < \infty$, namely (B.32) is satisfied, which leads in turn to the invariance property (B.28).

We have thus checked that the set $\mathfrak{M}_\lambda$ defined in (B.1) satisfies all the conditions required in Definition 4.2 in order to be a global random invariant Lipschitz manifold. The proof is now complete. □

References

1. L. Arnold, *Random Dynamical Systems*. Springer Monographs in Mathematics (Springer, Berlin, 1998)
2. L. Arnold, P. Imkeller, Normal forms for stochastic differential equations. Probab. Theory Relat. Fields **110**(4), 559–588 (1998)
3. L. Arnold, K. Xu, Normal forms for random differential equations. J. Differ. Equ. **116**(2), 484–503 (1995)
4. V.I. Arnold, *Geometrical Methods in the Theory of Ordinary Differential Equations*. Grundlehren der Mathematischen Wissenschaften, vol. 250 (Springer, New York, 1983)
5. B. Aulbach, A reduction principle for nonautonomous differential equations. Arch. Math. **39**, 217–232 (1982)
6. B. Aulbach, N. Van Minh, Nonlinear semigroups and the existence and stability of solutions of semilinear nonautonomous evolution equations. Abstr. Appl. Anal. **1**(4), 351–380 (1996)
7. B. Aulbach, T. Wanner, Integral manifolds for Carathéodory type differential equations in Banach spaces. In *Six Lectures on Dynamical Systems (Augsburg, 1994)* (World Scientific Publishing Co., River Edge, NJ, 1996) pp. 45–119
8. P.W. Bates, C.K.R.T. Jones, Invariant manifolds for semilinear partial differential equations. In *Dynamics Reported*, vol. 2. Dynam. Report. Ser. Dynam. Systems Appl., vol. 2. (Wiley, Chicester, 1989) pp. 1–38
9. P.W. Bates, K. Lu, C. Zeng, Existence and persistence of invariant manifolds for semiflows in Banach space. Mem. Amer. Math. Soc. **135**(645), viii+129 (1998)
10. P.W. Bates, K. Lu, C. Zeng, Approximately invariant manifolds and global dynamics of spike states. Invent. Math. **174**(2), 355–433 (2008)
11. J.R. Beddington, R.M. May, Harvesting natural populations in a randomly fluctuating environment. Science **197**(4302), 463–465 (1977)
12. A. Bensoussan, F. Flandoli, Stochastic inertial manifold. Stochast. Stochast. Rep. **53**, 13–39 (1995)
13. W.-J. Beyn, W. Kleß, Numerical Taylor expansions of invariant manifolds in large dynamical systems. Numer. Math. **80**(1), 1–38 (1998)
14. Y.N. Bibikov, *Local Theory of Nonlinear Analytic Ordinary Differential Equations*. Lecture Notes in Mathematics, vol. 702 (Springer, Berlin, 1979)
15. B. Birnir, *The Kolmogorov-Obukhov Theory of Turbulence: A Mathematical Theory of Turbulence*. Springer Briefs in Mathematics (Springer, New York, 2013)
16. D. Blömker, Amplitude Equations for Stochastic Partial Differential Equations. Interdisciplinary Mathematical Sciences, vol. 3 (World Scientific Publishing Co., Pte. Ltd., Hackensack, 2007)
17. D. Blömker, W.W. Mohammed, Amplitude equations for SPDEs with cubic nonlinearities. Stochastics **85**(2), 181–215 (2013)

M.D. Chekroun et al., *Approximation of Stochastic Invariant Manifolds*,
SpringerBriefs in Mathematics, DOI 10.1007/978-3-319-12496-4

18. D. Blömker, W. Wang, Qualitative properties of local random invariant manifolds for SPDEs with quadratic nonlinearity. J. Dyn. Diff. Equ. **22**, 677–695 (2010)
19. N.N. Bogoliubov, Y.A. Mitropolsky. *Asymptotic Methods in the Theory of Non-linear Oscillations*. Translated from the second revised Russian edition. International Monographs on Advanced Mathematics and Physics (Hindustan Publishing Corp., Delhi, Gordon and Breach Science Publishers, New York, 1961)
20. L. Boutet de Monvel, I.D. Chueshov, A.V. Rezounenko, Inertial manifolds for retarded semilinear parabolic equations. Nonlinear Anal. **34**(6), 907–925 (1998)
21. P. Boxler, A stochastic version of center manifold theory. Probab. Theory Relat. Fields **83**, 509–545 (1989)
22. P. Boxler, How to construct stochastic center manifolds on the level of vector fields. In *Lyapunov Exponents (Oberwolfach, 1990)*, Lecture Notes in Mathematics, vol. 1486 (Springer, Berlin, 1991) pp. 141–158
23. H.W. Broer, H.M. Osinga, G. Vegter, Algorithms for computing normally hyperbolic invariant manifolds. Z. Angew. Math. Phys. **48**(3), 480–524 (1997)
24. H.S. Brown, M.S. Jolly, I.G. Kevrekidis, E.S. Titi, Use of approximate inertial manifolds in bifurcation calculations. In *Continuation and Bifurcations: Numerical Techniques and Applications (Leuven, 1989)*, NATO Adv. Sci. Inst. Ser. C Math. Phys. Sci. vol. 313 (Kluwer Academic Publishers, Dordrecht, 1990) pp. 9–23
25. X. Cabré, E. Fontich, R. de la Llave, The parameterization method for invariant manifolds. III. Overview and applications. J. Differ. Equ. **218**(2), 444–515 (2005)
26. T. Caraballo, I. Chueshov, J.A. Langa, Existence of invariant manifolds for coupled parabolic and hyperbolic stochastic partial differential equations. Nonlinearity **18**(2), 747–767 (2005)
27. T. Caraballo, J. Duan, K. Lu, B. Schmalfuß, Invariant manifolds for random and stochastic partial differential equations. Adv. Nonlinear Stud. **10**(1), 23–52 (2010)
28. T. Caraballo, P.E. Kloeden, B. Schmalfuß, Exponentially stable stationary solutions for stochastic evolution equations and their perturbation. Appl. Math. Optim. **50**(3), 183–207 (2004)
29. T. Caraballo, J.A. Langa, J.C. Robinson, A stochastic pitchfork bifurcation in a reaction-diffusion equation. Proc. R. Soc. Lond. A **457**, 2041–2061 (2001)
30. J. Carr, *Applications of Centre Manifold Theory*. Applied Mathematical Sciences vol. 35 (Springer, New York, 1981)
31. A.N. Carvalho, J.A. Langa, J.C. Robinson, *Attractors for Infinite-Dimensional Non-autonomous Dynamical Systems*. Applied Mathematical Sciences, vol. 182 (Springer, New York, 2013)
32. C. Castaing, M. Valadier, *Convex Analysis and Measurable Multifunctions*. Lecture Notes in Mathematics, vol. 580 (Springer, Berlin, 1977)
33. T. Cazenave, A. Haraux, *An Introduction to Semilinear Evolution Equations*. Oxford Lecture Series in Mathematics and its Applications, vol. 13 (The Clarendon Press, Oxford University Press, New York, 1998)
34. M.D. Chekroun, H. Liu, Finite-horizon parameterizing manifolds, and applications to suboptimal control of nonlinear parabolic PDEs. *Acta Applicandae Mathematicae* (2014). http://dx.doi.org/10.1007/s10440-014-9949-1
35. M.D. Chekroun, H. Liu, S. Wang, Non-Markovian reduced systems for stochastic partial differential equations: The additive noise case. *Preprint* (2013). arXiv:1311.3069
36. M.D. Chekroun, H. Liu, S. Wang, Stochastic parameterizing manifolds: Application to stochastic transitions in SPDEs. *In Preparation* (2014)
37. M.D. Chekroun, H. Liu, S. Wang, *Stochastic Parameterizing Manifolds and Non-Markovian Reduced Equations: Stochastic Manifolds for Nonlinear SPDEs II*. Springer Briefs in Mathematics (Springer, New York, 2014) (to appear)
38. M.D. Chekroun, L.J. Roques, Models of population dynamics under the influence of external perturbations: mathematical results. C. R. Math. Acad. Sci. Paris **343**(5), 307–310 (2006)
39. M.D. Chekroun, E. Simonnet, M. Ghil, Stochastic climate dynamics: random attractors and time-dependent invariant measures. Physica D **240**(21), 1685–1700 (2011)

40. G. Chen, J. Duan, J. Zhang, Geometric shape of invariant manifolds for a class of stochastic partial differential equations. J. Math. Phys. **52**, 072702, 14 (2011)
41. X. Chen, A.J. Roberts, J. Duan, Center manifolds for stochastic evolution equations. *Preprint* (2012). arXiv:1210.5924
42. I.D. Cheushov, M. Scheutzow, Inertial manifolds and forms for stochastically perturbed retarded semilinear parabolic equations. J. Dyn. Diff. Equ. **13**(2), 355–380 (2001)
43. C. Chicone, Y. Latushkin, Center manifolds for infinite-dimensional nonautonomous differential equations. J. Differ. Equ. **141**, 356–399 (1997)
44. S.-N. Chow, J.K. Hale, *Methods of Bifurcation Theory*. Grundlehren der Mathematischen Wissenschaften, vol. 251 (Springer, New York, 1982)
45. S.-N. Chow, X.-B. Lin, K. Lu, Smooth invariant foliations in infinite-dimensional spaces. J. Differ. Equ. **94**(2), 266–291 (1991)
46. S.-N. Chow, K. Lu, Invariant manifolds for flows in Banach spaces. J. Differ. Equ. **74**, 285–317 (1988)
47. I. Chueshov, *Monotone Random Systems: Theory and Applications*. Lecture Notes in Mathematics, vol. 1779 (Springer, Berlin, 2002)
48. I. Chueshov, A. Millet, Stochastic 2D hydrodynamical type systems: well posedness and large deviations. Appl. Math. Optim. **61**(3), 379–420 (2010)
49. I.D. Chueshov, Global attractors for non-linear problems of mathematical physics. Russ. Math. Surv. **48**(3), 133–161 (1993)
50. I.D. Chueshov, Approximate inertial manifolds of exponential order for semilinear parabolic equations subjected to additive white noise. J. Dyn. Diff. Equ. **7**(4), 549–566 (1995)
51. I.D. Chueshov, T.V. Girya, Inertial manifolds and forms for semilinear parabolic equations subjected to additive white noise. Lett. Math. Phys. **34**, 69–76 (1995)
52. P. Constantin, C. Foias, B. Nicolaenko, R. Temam, *Integral Manifolds and Inertial Manifolds for Dissipative Partial Differential Equations*. Applied Mathematical Sciences, vol. 70 (Springer, New York, 1989)
53. P.H. Coullet, C. Elphick, E. Tirapegui, Normal form of a Hopf bifurcation with noise. Phys. Lett. A **111**(6), 277–282 (1985)
54. H. Crauel, *Random Probability Measures on Polish Spaces* (Taylor & Francis Inc., London, 2002)
55. H. Crauel, F. Flandoli, Attractors for random dynamical systems. Probab. Theory Relat. Fields **100**, 365–393 (1994)
56. M.C. Cross, P.C. Hohenberg, Pattern formation outside of equilibrium. Rev. Mod. Phys. **65**(3), 851–1112 (1993)
57. G. Da Prato, A. Debussche, Construction of stochastic inertial manifolds using backward integration. Stochast. Stochast. Rep. **59**(3–4), 305–324 (1996)
58. G. Da Prato, J. Zabczyk, *Stochastic Equations in Infinite Dimensions*. Encyclopedia of Mathematics and its Applications, vol. 44 (Cambridge University Press, Cambridge, 2008)
59. A. Dávid, S.C. Sinha, Bifurcation analysis of nonlinear dynamic systems with time-periodic coefficients. In *Bifurcation and Chaos in Complex Systems*. Edited Series on Advances in Nonlinear Science and Complexity, vol. 1 (Elsevier, Burlington, 2006) pp. 279–338
60. A. Debussche, R. Temam, Convergent families of approximate inertial manifolds. J. Math. Pure Appl. **73**(5), 489–522 (1994)
61. A. Debussche, R. Temam, Inertial manifolds with delay. Appl. Math. Lett. **8**(2), 21–24 (1995)
62. A. Debussche, R. Temam, Some new generalizations of inertial manifolds. Discrete Contin. Dyn. Syst. **2**, 543–558 (1996)
63. C. Devulder, M. Marion, E.S. Titi, On the rate of convergence of the nonlinear Galerkin methods. Math. Comp. **60**(202), 495–514 (1993)
64. M.P. do Carmo, *Differential Geometry of Curves and Surfaces* (Prentice-Hall Inc., Englewood Cliffs, 1976)
65. J. Duan, K. Lu, B. Schmalfuss, Invariant manifolds for stochastic partial differential equations. Ann. Probab. **31**, 2109–2135 (2003)

66. J. Duan, K. Lu, B. Schmalfuss, Smooth stable and unstable manifolds for stochastic evolutionary equations. J. Dyn. Diff. Equ. **16**, 949–972 (2004)
67. J. Duan, W. Wang, *Effective Dynamics of Stochastic Partial Differential Equations* (Elsevier, Amsterdam, 2014)
68. R.M. Dudley, *Real Analysis and Probability*. Cambridge Studies in Advanced Mathematics, vol. 74 (Cambridge University Press, Cambridge, 2002)
69. W. E, J.C. Mattingly, Ya. Sinai, Gibbsian dynamics and ergodicity for the stochastically forced Navier-Stokes equation. Commun. Math. Phys. **224**(1), 83–106 (2001)
70. T. Eirola, J. von Pfaler, Numerical Taylor expansions for invariant manifolds. Numer. Math. **99**(1), 25–46 (2004)
71. K.-J. Engel, R. Nagel, *One-Parameter Semigroups for Linear Evolution Equations*. Graduate Texts in Mathematics, vol. 194 (Springer, New York, 2000)
72. L.C. Evans, *Partial Differential Equations*. Graduate Studies in Mathematics, vol. 19 (American Mathematical Society, Providence, 2010)
73. L.C. Evans, *An Introduction to Stochastic Differential Equations* (American Mathematical Society, Providence, 2013)
74. T. Faria, Normal forms and Hopf bifurcation for partial differential equations with delays. Trans. Amer. Math. Soc. **352**(5), 2217–2238 (2000)
75. T. Faria, Normal forms for semilinear functional differential equations in Banach spaces and applications Part II. Discrete Contin. Dyn. Syst. **7**(1), 155–176 (2001)
76. F. Flandoli. An introduction to 3D stochastic fluid dynamics. In *SPDE in Hydrodynamic: Recent Progress and Prospects*. Lecture Notes in Mathematics, vol. 1942 (Springer, Berlin, 2008) pp. 51–150
77. F. Flandoli, H. Lisei, Stationary conjugation of flows for parabolic SPDEs with multiplicative noise and some applications. Stochast. Anal. Appl. **22**(6), 1385–1420 (2004)
78. C. Foias, O. Manley, R. Temam, Modelling of the interaction of small and large eddies in two-dimensional turbulent flows. RAIRO Modél Math. Anal. Numér. **22**(1), 93–118 (1988)
79. C. Foias, G.R. Sell, R. Temam, Inertial manifolds for nonlinear evolutionary equations. J. Differ. Equ. **73**(2), 309–353 (1988)
80. C. Foias, G.R. Sell, E.S. Titi, Exponential tracking and the approximation of inertial manifolds for dissipative nonlinear equations. J. Dyn. Differ. Equ. **1**, 199–244 (1989)
81. C. Foias, R. Temam, Approximation of attractors by algebraic or analytic sets. SIAM J. Math. Anal. **25**(5), 1269–1302 (1994)
82. G.B. Folland, *Real Analysis: Modern Techniques and Their Applications* (Wiley, New York, 1999)
83. E. Forgoston, I.B. Schwartz, Escape rates in a stochastic environment with multiple scales. SIAM J. Appl. Dyn. Syst. **8**(3), 1190–1217 (2009)
84. T. Gallay, A center-stable manifold theorem for differential equations in Banach spaces. Commun. Math. Phys. **152**(2), 249–268 (1993)
85. B. Gess, Random attractors for degenerate stochastic partial differential equations. J. Dyn. Differ. Equ. **25**, 121–157 (2013)
86. M. Hairer, Ergodic properties of a class of non-Markovian processes. In *Trends in Stochastic Analysis*. London Mathematical Society Lecture Note Series, vol. 353 (Cambridge University Press, Cambridge, 2009)
87. M. Hairer, A. Ohashi, Ergodic theory for SDEs with extrinsic memory. Ann. Probab. **35**, 1950–1977 (2007)
88. J.K. Hale, *Ordinary Differ. Equ.*, 2nd edn. (Robert E. Krieger Publishing Co., Huntington, 1980)
89. J.K. Hale, *Asymptotic Behavior of Dissipative Systems*. Mathematical Surveys and Monographs, vol. 25 (American Mathematical Society, Providence, 1988)
90. M. Haragus, G. Iooss, *Local Bifurcations, Center Manifolds, and Normal Forms in Infinite-Dimensional Dynamical Systems*. Universitext (Springer, London, 2011)
91. A. Haro, Automatic differentiation methods in computational dynamical systems: Invariant manifolds and normal forms of vector fields at fixed points, *Universidad de Barcelona, Preprint* (2008)

92. D. Henry, *Geometric Theory of Semilinear Parabolic Equations*. Lecture Notes in Mathematics, vol. 840 (Springer, Berlin, 1981)
93. G. Hetzer, W. Shen, S. Zhu, Asymptotic behavior of positive solutions of random and stochastic parabolic equations of Fisher and Kolmogorov types. J. Dyn. Differ. Equ. **14**(1), 139–188 (2002)
94. M.W. Hirsch, C.C. Pugh, M. Shub, *Invariant Manifolds*. Lecture Notes in Mathematics, vol. 583 (Springer, Berlin, 1977)
95. P. Imkeller, C. Lederer, On the cohomology of flows of stochastic and random differential equations. Probab. Theory Relat. Fields **120**(2), 209–235 (2001)
96. P. Imkeller, C. Lederer, The cohomology of stochastic and random differential equations, and local linearization of stochastic flows. Stochast. Dyn. **2**(02), 131–159 (2002)
97. P. Imkeller, B. Schmalfuss, The conjugacy of stochastic and random differential equations and the existence of global attractors. J. Dyn. Differ. Equ. **13**(2), 215–249 (2001)
98. M.E. Johnson, M.S. Jolly, I.G. Kevrekidis, Two-dimensional invariant manifolds and global bifurcations: some approximation and visualization studies. Numer. Algorithms **14**(1–3), 125–140 (1997)
99. M.S. Jolly, I.G. Kevrekidis, E.S. Titi, Approximate inertial manifolds for the Kuramoto-Sivashinsky equation: analysis and computations. Physica D **44**(1), 38–60 (1990)
100. M.S. Jolly, R. Rosa, Computation of non-smooth local centre manifolds. IMA J. Numer. Anal. **25**, 698–725 (2005)
101. M.S. Jolly, R. Rosa, R. Temam, Accurate computations on inertial manifolds. SIAM J. Sci. Comput. **22**(6), 2216–2238 (2001)
102. D.A. Jones, E.S. Titi, A remark on quasi-stationary approximate inertial manifolds for the Navier-Stokes equations. SIAM J. Math. Anal. **25**(3), 894–914 (1994)
103. X. Kan, J. Duan, I.G. Kevrekidis, A.J. Roberts, Simulating stochastic inertial manifolds by a backward-forward approach. SIAM J. Appl. Dyn. Syst. **12**(1), 487–514 (2013)
104. T. Kato, *Perturbation Theory for Linear Operators*. Classics in Mathematics (Springer, Berlin, 1995)
105. A. Kelley, The stable, center-stable, center, center-unstable, unstable manifolds. J. Differ. Equ. **3**, 546–570 (1967)
106. P.E. Kloeden, J.A. Langa, Flattening, squeezing and the existence of random attractors. Proc. R. Soc. Lond. Ser. A **463**(2077), 163–181 (2007)
107. H. Koch, On center manifolds. Nonlinear Anal. **28**(7), 1227–1248 (1997)
108. B. Krauskopf, H.M. Osinga, E.J. Doedel, M.E. Henderson, J. Guckenheimer, A. Vladimirsky, M. Dellnitz, O. Junge, A survey of methods for computing (un)stable manifolds of vector fields. Int. J. Bifurcat. Chaos **15**(03), 763–791 (2005)
109. N. Krylov, N.N. Bogoliubov, *The Application of Methods of Nonlinear Mechanics to the Theory of Stationary Oscillations*. Publication No. 8 of Ukrainian Academy of Science, Kiev, U.S.S.R. (1934) (in Russian)
110. H. Kunita, *Stochastic Flows and Stochastic Differential Equations*. Cambridge studies in advanced mathematics, vol. 24 (Cambridge University Press, Cambridge, 1990)
111. Y.A. Kuznetsov, *Elements of Applied Bifurcation Theory*. Applied Mathematical Sciences, vol. 112, 3rd ed. (Springer, New York, 2004)
112. S. Lang, *Real and Functional Analysis*. Graduate Texts in Mathematics, vol. 142, 3rd ed. (Springer, New York, 1993)
113. Z. Lian, K. Lu, Lyapunov exponents and invariant manifolds for random dynamical systems in a Banach space. Mem. Amer. Math. Soc. **206**(967), vi+106 (2010)
114. A. Liapounoff, *Problème Général de la Stabilité du Mouvement*. Annals of Mathematics Studies, vol. 17 (Princeton University Press, Princeton, 1947)
115. W. Liu, Large deviations for stochastic evolution equations with small multiplicative noise. Appl. Math. Optim. **61**(1), 27–56 (2010)
116. Y. Lv, A.J. Roberts, Averaging approximation to singularly perturbed nonlinear stochastic wave equations. J. Math. Phys. **53**, 062702, 11 (2012)

117. T. Ma, S. Wang, *Bifurcation Theory and Applications*. World Scientific Series on Nonlinear Science. Series A: Monographs and Treatises, vol. 53 (World Scientific Publishing Co., Pte. Ltd., Hackensack, 2005)
118. T. Ma, S. Wang, Dynamic transition theory for thermohaline circulation. Physica D **239**, 167–189 (2010)
119. T. Ma, S. Wang, El niño southern oscillation as sporadic oscillations between metastable states. Adv. Atmos. Sci. **28**, 612–622 (2011)
120. T. Ma, S. Wang, *Phase Transition Dynamics* (Springer, New York, 2013)
121. P. Magal, S. Ruan, Center manifolds for semilinear equations with non-dense domain and applications to Hopf bifurcation in age structured models. Mem. Amer. Math. Soc. **202**(951), vi+71 (2009)
122. M. Marion, R. Temam, Nonlinear Galerkin methods. SIAM J. Numer. Anal. **26**(5), 1139–1157 (1989)
123. S.-E.A. Mohammed, M.K.R. Scheutzow, The stable manifold theorem for stochastic differential equations. Ann. Probab. **27**(2), 615–652 (1999)
124. S.-E.A. Mohammed, T. Zhang, H. Zhao, The stable manifold theorem for semilinear stochastic evolution equations and stochastic partial differential equations. Mem. Amer. Math. Soc. **196**(917), vi+105 (2008)
125. C. Mueller, L. Mytnik, J. Quastel, Small noise asymptotics of traveling waves. Markov Process Relat. Fields **14**(3), 333–342 (2008)
126. C. Mueller, L. Mytnik, J. Quastel, Effect of noise on front propagation in reaction-diffusion equations of KPP type. Inventiones Math. **184**(2), 405–453 (2011)
127. M.A. Muñoz, Multiplicative noise in non-equilibrium phase transitions: a tutorial. In *Advances in Condensed Matter and Statistical Physics* (Nova Science Publishers Inc, Commack, 2004) pp. 37–68
128. N.S. Namachchivaya, Y.K. Lin, Method of stochastic normal forms. Int. J. Non-linear Mech. **26**(6), 931–943 (1991)
129. J. Novo, E.S. Titi, S. Wynne, Efficient methods using high accuracy approximate inertial manifolds. Numer. Math. **87**(3), 523–554 (2001)
130. B. Øksendal, *Stochastic Differential Equations: An Introduction with Applications*, 5th edn. (Springer, New York, 1998)
131. O. Perron, Über stabilität und asymptotisches verhalten der integrale von differentialgleichungssystemen. Math. Z. **29**(1), 129–160 (1929)
132. V.A. Pliss, The reduction principle in the theory of stability of motion. Sov. Math. Dokl. **5**, 247–250 (1964)
133. C. Pötzsche, Bifurcations in nonautonomous dynamical systems: Results and tools in discrete time. In *Proceedings of the Workshop Future Directions in Difference Equations*, Coleccion Congress, vol. 69 (University of Vigo, Service Publication, Vigo, 2011) pp. 163–212
134. C. Pötzsche, M. Rasmussen, Taylor approximation of integral manifolds. J. Dyn. Differ. Equ. **18**(2), 427–460 (2006)
135. C. Pötzsche, M. Rasmussen, Computation of integral manifolds for Carathéodory differential equations. IMA J. Numer. Anal. **30**(2), 401–430 (2010)
136. A.J. Roberts, Normal form transforms separate slow and fast modes in stochastic dynamical systems. Physica A **387**, 12–38 (2008)
137. J.C. Robinson, The asymptotic completeness of inertial manifolds. Nonlinearity **9**(5), 1325–1340 (1996)
138. J.C. Robinson, *Infinite-Dimensional Dynamical Systems: An Introduction to Dissipative Parabolic PDEs and the Theory of Global Attractors*. Cambridge Texts in Applied Mathematics (Cambridge University Press, Cambridge, 2001)
139. L. Roques, M.D. Chekroun, On population resilience to external perturbations. SIAM J. Appl. Math. **68**(1), 133–153 (2007)
140. L. Roques, M.D. Chekroun, Does reaction-diffusion support the duality of fragmentation effect? Ecol. Complex. **7**(1), 100–106 (2010)

141. D. Ruelle, Characteristic exponents and invariant manifolds in Hilbert space. Ann. Math. **115**, 243–290 (1982)
142. B. Schmalfuss, Inertial manifolds for random differential equations. In *Probability and Partial Differential Equations in Modern Applied Mathematics*. IMA Volumes—Institute for Mathematics and its Applications, vol. 140 (Springer, New York, 2005) pp. 213–236
143. G. Sell, Y. You, *Dynamics of Evolutionary Equations*. Applied Mathematical Sciences, vol. 143 (Springer, New York, 2002)
144. A. Shirikyan, S. Zelik, Exponential attractors for random dynamical systems and applications. Stoch. PDE: Anal. Comp. **1**, 241–281 (2013)
145. X. Sun, J. Duan, X. Li, An impact of noise on invariant manifolds in nonlinear dynamical systems. J. Math. Phys. **51**, 042702, 12 (2010)
146. A.E. Taylor, *Introduction to Functional Analysis* (Wiley, New York, 1958)
147. R. Temam, *Infinite-Dimensional Dynamical Systems in Mechanics and Physics*. Applied Mathematical Sciences, vol. 68, 2nd ed. (Springer, New York, 1997)
148. R. Temam, D. Wirosoetisno, Stability of the slow manifold in the primitive equations. SIAM J. Math. Anal. **42**(1), 427–458 (2010)
149. R. Temam, D. Wirosoetisno, Slow manifolds and invariant sets of the primitive equations. J. Atmos. Sci. **68**(3), 675–682 (2011)
150. B. Texier, K. Zumbrun, Galloping instability of viscous shock waves. Physica D **237**(10), 1553–1601 (2008)
151. E.S. Titi, On approximate inertial manifolds to the Navier-Stokes equations. J. Math. Anal. Appl. **149**(2), 540–557 (1990)
152. A. Vanderbauwhede, Centre manifolds, normal forms and elementary bifurcations. In *Dynamics Reported*, vol. 2. Dynam. Report. Ser. Dynam. Systems Appl., vol. 2 (Wiley, Chichester, 1989) pp. 89–169
153. A. Vanderbauwhede, G. Iooss, Center manifold theory in infinite dimensions. In *Dynamics Reported: Expositions in Dynamical Systems*. Dynam. Report. Expositions Dynam. Systems (N.S.), vol. 1 (Springer, Berlin, 1992) pp. 125–163
154. A. Vanderbauwhede, S.A. Van Gils, Center manifolds and contractions on a scale of Banach spaces. J. Funct. Anal. **72**(2), 209–224 (1987)
155. W. Wang, J. Duan. A dynamical approximation for stochastic partial differential equations. J. Math. Phys. **48**, 102701, 14 (2007)
156. C. Xu, A.J. Roberts, On the low-dimensional modelling of Stratonovich stochastic differential equations. Physica A **225**(1), 62–80 (1996)

Index

M.D. Chekroun et al., *Approximation of Stochastic Invariant Manifolds*,
SpringerBriefs in Mathematics, DOI 10.1007/978-3-319-12496-4